Contents

Illustration Credits
Oldriska Ceska did the drawings reproduced on pages 41, 48, 73, 85, 93, 96, 110, 117, 134 (left), 138, 142, 147, 154, and 158; Kungold Hillis did those on pages 49, 50, 56, 57, 68, 69, 80, 81, and 134 (right).

T.C. Brayshaw took the colour photograph reproduced on page 152; Jim Pojar took the one reproduced on page 101 and the top one on page 122. The photograph on page 114 was taken by Janet Renfroe. Robert D. Turner and Nancy J. Turner took the cover photograph, and the photographs on pages 30, 54, 62, 66, 78, 90, 104, and 127 as well as the bottom one on page 122.

Wild Green Vegetables of Canada

Published by the
National Museums of Canada

Coordination
Madeleine Choquette-Delvaux

Editor
Penny Williams

Production
Donald Matheson

Series Design Concept
Eskind Waddell

Layout Design
Gregory Gregory Limited

Typesetting
The Runge Press Limited

Printing
K.G. Campbell Corporation

Publications in the
Edible Wild Plants of Canada series:

1. *Edible Garden Weeds of Canada*, 1978
2. *Wild Coffee and Tea Substitutes of Canada*, 1978
3. *Edible Wild Fruits and Nuts of Canada*, 1979
4. *Wild Green Vegetables of Canada*, 1980

Cette collection existe en français
sous le titre, Plantes sauvages
comestibles du Canada:

1. *Mauvaises herbes comestibles de nos jardins*, 1978
2. *Succédanés sauvages du thé et du café au Canada*, 1978
3. *Fruits et noix sauvages comestibles du Canada*, 1979
4. *Légumes sauvages du Canada*, 1980

Cover: nodding onions (*Allium cernuum*)

Adam F. Szczawinski
Nancy J. Turner

Edible Wild Plants
of Canada, No. 4

Wild Green Vegetables of Canada

National Museum
of Natural Sciences

National Museums
of Canada

National Museum of Natural Sciences
National Museums of Canada
Ottawa, Canada K1A 0M8

Catalogue No. NM95-40/4

Distributed in the United States,
Central and South America,
Australia, New Zealand and
Southeast Asia by:
The University of Chicago Press
5801 South Ellis Avenue
Chicago, Illinois 60637

Printed in Canada

English Edition
ISBN 0-660-10342-7
ISSN 0705-3967

French Edition
ISBN 0-660-90256-7
ISSN 0705-3975

Acknowledgements

We should like to acknowledge the continuing advice and support of the staff of the National Museums of Canada, and especially Mr. Irwin M. Brodo, Curator of Lichens, and Dr. J. H. Soper, Chief Botanist, National Museum of Natural Sciences. We are grateful to the following people for their suggestions and contributions to the book: Dr. Robert T. Ogilvie, Curator of Botany, British Columbia Provincial Museum, Victoria; Mrs. Mary I. Moore, Deep River, Ontario; Dr. Douglas Leechman, formerly with the Museum of Man, Ottawa, now retired in Victoria; Mr. Bill Cody, Experimental Farm, Agriculture Canada, Ottawa.

We should also like to thank those people who provided recipes and specific information on wild green vegetables: Mrs. Constance Conrader of Oconomowoc, Wisconsin; Dr. Judith Madlener of Berkeley, California for permission to use her recipes from *The Seavegetable Book*; Enid K. Lemon and Lissa Calvert of Victoria, B.C. whose recipe is taken from their collection of printed notepaper, "Pick'n'Cook Notes"; and Mrs. Joy Inglis, Quathiaski Cove, B.C.

Finally, we should particularly like to thank Dr. Tina Kuiper-Goodman, Toxicologist with the Bureau of Chemical Safety, Toxicological Evaluation Division, Department of Health and Welfare, Ottawa, for her numerous suggestions and valuable information on the potential toxicology of the plants included in this book.

Preface

The culinary possibilities of wild green vegetables have been exploited for centuries in Europe and the Orient, but we North Americans are notoriously suspicious of any plant collected in its natural habitat. We let the reputation of a few poisonous or unpalatable species deprive us of the enjoyment to be found in others that are both safe and delicious. Many people lack confidence in their ability to identify wild plants correctly, but it does not take much botanical training to become acquainted with a dozen or so good edible species and to learn to recognize them on sight, just as one learns to recognize garden vegetables.

Many wild greens are as tasty and nutritious as those from your own garden or a grocery store. Some are excellent as salad ingredients, others are better when lightly cooked. Almost all can be used in soups, stews and casseroles, alone, mixed with other wild green vegetables, or combined with garden vegetables.

Non-cultivated green vegetables can be found almost anywhere in Canada. Some, especially those introduced from Europe and other parts of the world, are quite weedy and are common in and around gardens and other disturbed sites near human settlements. These have already been discussed in detail in our first book of this series, *Edible Garden Weeds of Canada*, but are also listed here in a table (see pages 23–26). However, most of this volume deals with wild green vegetables—those that are native to Canada and found mainly in their natural habitats in woodlands, meadows, seashores, oceans, marshes, and alpine and Arctic tundra zones.

More than twenty-five species or groups of related species are described here in detail. For each, we have provided a botanical description, notes on habitat and distribution in Canada, and general information on how it can be used as food. Also provided are specific recipes, and additional notes of interest on the history and folklore of the plants, as well as their uses in other countries and cultures, by native peoples, or their use as materials, dyes, or medicines. All the plants are illustrated by colour photographs or black and white drawings. For purposes of comparison, we have provided drawings of certain poisonous species that could be confused with some of the edible species included in this book.

Some edible species, including marine algae, cat-tail, fireweed, wild onions, and cow parsnip, are very common, easily gathered, and delicious when properly prepared. Others, such as cactus, mountain bistorts, woolly lousewort, and rock tripe lichen, are either restricted in distribution and frequency or are of marginal edibility, but are included because of their value as survival foods.

Historically, wild greens have played a significant role in the development of our country. Explorers and mariners of the eighteenth and nineteenth centuries soon learned the value of using fresh wild greens, such as wild onions, scurvy-grass, seaside plantain, and fireweed, to supplement their food stores and to provide the vitamins and minerals necessary to prevent scurvy and other diseases related to nutritional deficiency. Indians and Inuit peoples also were aware of the nutritive qualities of wild green vegetables and relied on them, especially in springtime when most greens are at their best, to provide a welcome change from the meagre wintertime fare of dried and preserved foods. Green vegetables eaten by native peoples include marine algae, eel-grass, cow parsnip, "Indian celery", rock-cress, prickly-pear cactus, mountain sorrel, thimbleberry and salmonberry shoots, and woolly lousewort.

Some of these greens, or their relatives, have been and still are widely used as vegetables in other parts of the world. For example, marine algae are highly regarded by Oriental peoples and are also popular in parts of the United Kingdom and northern Europe. Stonecrop, bistort, glasswort, and scurvy-grass are still eaten in parts of the United Kingdom and Europe. Seabeach-sandwort is used in Iceland to make a type of sauerkraut, and cat-tail shoots are eaten by the Cossack peoples of the U.S.S.R.

Canadians could learn much from their own past and from other peoples of the world about the potential of wild green vegetables. Campers, wayside wanderers, and wilderness backpackers could use them to supplement or replace their regular vegetable reserves. For those people living in remote areas of the country where fresh garden produce is scarce, expensive, and often of poor quality, wild greens could become a cheap and satisfying alternative to commercially grown vegetables. After all, it makes much more sense to utilize wild plants already well adapted to survival in the harsh climates of the North or along the rugged seacoasts than to attempt to cultivate conventional vegetables basically unsuited to such environments, or to import vegetables from long distances at great trouble and expense.

Aside from being pleasurable and economically expedient, using wild greens can, under some circumstances, save your life. During the Second World War, in the winter of 1939, Adam Szczawinski was forced to flee from Poland and had to survive for five days in the high alpine regions of the Carpathian Mountains. In desperation, he turned to wild plants, such as lichens, for survival. This incident taught him not only to respect the value of plants but also instilled in him the desire to learn more about edible species and to teach others about their potential as aids to survival.

Increased oil and mineral exploration and development in remote areas, wider recreational activities in the wilderness, and an ever-expanding system of commercial air and ground travel have all led to a greater possibility of people becoming lost or stranded in areas far from civilization. Everyone should have at least minimal survival training including some knowledge of the most important and common wild edible plants.

Editor's Note

The transition from the imperial system of weights and measures to the international metric system has prompted the use of both systems in this series on edible wild plants. Most of the recipes were originally developed under the imperial system, and have been converted to their closest metric equivalents based on the Canadian Metric Commission's guidelines. These "equivalents" are actually replacements, and sometimes involve slightly larger amounts than the original imperial measures. For example, the 8-ounce cup is actually equivalent to 225 mL, but the standard metric measure that replaces it holds 250 mL. As a result, the metric conversions of the recipes sometimes yield slightly larger amounts than do the original recipes.

Readers of this series will notice the change in the symbol for millilitres from ml in the first two publications, to mL in the third publication, and the present one. The change was prompted by the decision of the Metric Commission to adopt mL as the official symbol.

Wild Greens as Food

"Eat your vegetables!" How many of us have heard these words as children and complied with the command only under protest? Fortunately, the prospect of eating green vegetables seems to become more bearable as one grows older, and most adults recognize not only the nutritional importance of greens but actually enjoy eating them, or at least some of them. With the selection of available vegetables widened by the addition of wild greens, people with even the most finicky of tastes should find a vegetable to their liking.

Many wild green vegetables are comparable in taste and quality to domesticated ones, and in our opinion, some are even better. As a group, cultivated green vegetables are relatively low in usable protein and carbohydrates, but are an important source of many vitamins, especially vitamins A, B (including thiamin, riboflavin, folic acid, niacin), C and K, and minerals (including calcium, phosphorus, potassium, iron, chlorine, cobalt, and manganese). In addition, they provide fibre, or bulk, to our diet. This consists mainly of cellulose, which is indigestible but essential for proper functioning of the human digestive system.

Very few analyses have been carried out on the nutritional content in wild green vegetables, but available evidence indicates that they are as nutritious as cultivated ones. Certainly in the past they played the same role in the diets of Indian and Inuit peoples as domesticated greens do in our present-day diets. In his book *Stalking the Healthful Herbs*, Euell Gibbons provides comparative nutritional information for some wild and garden vegetables, but most of the wild greens he mentions were already discussed in our first book of this series, *Edible Garden Weeds of Canada*. Many of those included in this volume have not been analyzed. However, one study of note, by Hoffman *et al.*, did analyze the summertime ascorbic acid (vitamin C) and carotene (vitamin A) content of some of the species included here, along with many other native eastern Arctic plants. The results showed that the leaves of fireweed (*Epilobium angustifolium*) and mountain bistort (*Polygonum viviparum*), and the seedpods and stems of scurvy-grass (*Cochlearia officinalis*) all contained over 100 mg of ascorbic acid per 100 g fresh weight (fireweed contained 220 mg in one sample). The leaves of roseroot (*Sedum roseum*) and seabeach-sandwort (*Arenaria peploides*), and the leaves and stems of mountain sorrel (*Oxyria digyna*) contained lesser, but still significant, amounts ranging from 40 mg to 68 mg per 100 g fresh weight. Fireweed leaves were also high in β-carotene (112 μg per g in the two samples analyzed), and mountain sorrel, seabeach-sandwort, scurvy-grass, and roseroot contained lesser amounts ranging from 25 μg to 53 μg per g. (Mountain bistort leaves were apparently not analyzed for carotene content.)

One type of wild greens, the sea vegetables, or seaweeds, deserves special recognition for its nutritional qualities. In fact, the nutritive values of some sea vegetables exceed those of any other food source known to man, according to Judith Madlener in *The Seavegetable Book*. Many sea vegetables are high in digestible proteins, carbohydrates, fats and oils, vitamins, and minerals. The nutritive values of sea vegetables are discussed further on page 34.

In selecting the wild green vegetables to be included in this book, we considered several factors. We tried to choose species that gave the greatest culinary possibilities—those that are most easily gathered and best tasting. Many are favourites with us; some of them, such as wild onions and "Indian celery", we even cultivate in our gardens. We also made an effort to include plants from a variety of habitats and from the different geographical regions of Canada. Naturally, none of the greens is common to all of Canada but most people should be able to find at least some of them locally. Particular attention was paid to wild greens occurring in the Far North and along the seacoasts because we felt that people living in these areas, where domesticated vegetables, if available at all, are difficult to grow and expensive to buy, will be able to benefit most from wild greens. Species that have particular historical significance as food for native peoples or for explorers, mariners, and early settlers, were also given particular consideration.

The species described here do not by any means represent an exhaustive list of edible wild greens. In some cases, we simply did not have enough first-hand information on a species reported to be edible. Some of the "edible" plants that we did sample we found unsatisfactory and could not recommend them.

In a few cases, species could not be included because of serious potential health risks involved in their use. Pokeweed (*Phytolacca americana* L.), whose young leafy shoots have long been a favourite spring potherb in eastern North America, is a prime example. The mature foliage and fruits are highly toxic, due to the presence of a group of chemicals (triterpine saponins), but these compounds are not present in any quantity in the young growth and are destroyed by proper cooking. Apart from this danger, however, there is a much more serious problem. Pokeweed plants possess as yet unidentified mitogens, compounds that can be absorbed through any skin abrasions or cuts and that cause serious blood aberrations, according to Lewis and Elwin-Lewis in their recent book, *Medical Botany*. For these reasons, *Phytolacca* species should be handled only with extreme caution (make sure to wear gloves), and no matter how tasty, should not be eaten.

Another example of a widely enjoyed, but potentially dangerous green is bracken fern [*Pteridium aquilinum* (L.) Kuhn.] at the young fiddlehead stage. The toxic properties of mature bracken, due to the presence of the enzyme, thiaminase, have long been known, but recent evidence shows that bracken plants, even at the fiddlehead stage, contain carcinogenic substances and therefore can no longer be recommended as edible.

We have tried to present a balanced selection of edible greens, including many different types of plants and a variety of recipes, ranging from salads, sandwiches, and pickles to soups and casseroles. In addition, most of the green vegetables described are important survival foods, especially when used in combination with other types of wild edibles.

A Necessary Caution

The dangerous properties of pokeweed greens and bracken fiddleheads have already been mentioned. They can no longer be recommended, though undoubtedly people will continue eating them, as they have done for centuries.

It cannot be stressed enough that some plants can be harmful and even deadly. Even domesticated species, used the world over, are potentially dangerous. The potato is a good example. The plants of potato and its relative, the tomato, contain a compound called solanine that is violently toxic when taken in concentrated doses. Potato and tomato vines have killed livestock on many occasions. Solanine is also present in potato tubers (the edible portion) but fortunately usually not in sufficient quantities to cause illness. Kingsbury, in his book *Deadly Harvest*, states that the normal solanine content in potato tubers is about one-tenth the amount that would produce symptoms of poisoning. One would have to eat ten large meals of potatoes all at once before feeling any toxic effects. However, solanine tends to become concentrated in the sprouts and in the layer next to the skin of the tuber, particularly when the tuber has been exposed to the sunlight and has turned green, and if these portions are consumed raw in quantity, they can be lethal.

Lima beans contain a cyanogenic glycoside (a compound that yields cyanide) that can be present at toxic levels, especially in beans grown in tropical areas. Lima beans grown in North America yield only very small amounts of cyanide (less than 0.02 per cent) and are not toxic. Onions have also caused poisoning of livestock when consumed in large or even moderate amounts. Many members of the mustard family— cabbage, broccoli, turnip, and rutabaga, for example—may contain chemicals that promote goitre by interfering with thyroid hormone formation. Spinach and beet tops contain soluble salts of oxalic acid that decrease the body's calcium retention, and consumption of very large quantities with a low calcium diet has led to death in laboratory animals.

These examples are given to illustrate the point that almost any food, taken in excessive doses over an extended period of time, and in excessive concentrations to the exclusion of other foods, can be harmful. Wild foods, including wild greens, must be treated in the same rational way that domesticated foods are. If you are trying them for the first time, take only a small quantity. Even when you are sure they cause you no problems, it is a good idea to be moderate in your intake and eat as varied a diet as possible.

Be positive of the identity of the plants you eat. The experimental approach to eating wild greens may be interesting but can be deadly. There are not many wild plants in Canada that are so poisonous as to have immediate and fatal toxic effects with the ingestion of small quantities, but such plants do exist. One of these is water hemlock (*Cicuta* species), a genus in the celery family. Members of this genus have hollow stalks, small white flowers in numerous umbels, pinnately-compound, finely dissected leaves up to 1 m long when mature, and clustered, tuberous roots with a swollen, chambered, turnip-like rootstock. When cut, the rootstock exudes yellowish, oily drops with the characteristic strong, pungent odour of raw parsnip. The plants usually grow in standing water and are found in marshes, ditches, swamps, and around lake edges. The genus *Cicuta* is considered by many authorities to be the most violently poisonous plant of the North Temperate Zone, according to Kingsbury in *Poisonous Plants of the United States and Canada*. The entire plant is poisonous, the rootstock being the most virulent. The poison is a complex, highly unsaturated higher alcohol known as cicutoxin. It acts directly on the central nervous system, and a single root can easily kill a cow. Death can occur within half an hour after consumption.

Other potentially deadly, rapidly acting species include false, or Indian, hellebore (*Veratrum viride* Ait.) and death, or white, camas (*Zygadenus*, or *Zigadenus* species, see illustration, page 57). Both these are in the lily family and in both cases, their toxicity is due, at least in part, to the presence of a number of steroid alkaloids, described in detail by Kingsbury. Indian hellebore is an erect, coarse, perennial herb with stems rising from a short, thickened rootstock. The stems, up to 2 m tall, are unbranched and leafy. The leaves are alternate, broadly oval, or elliptical, up to 25 cm long and about half as broad, decreasing in size towards the top of the plant. They are conspicuously parallel-veined and sharply pleated. The flowers are greenish and numerous, in a large, terminal, somewhat drooping cluster. This plant occurs in moist grassy meadows and swampy ground in British Columbia and southwestern Alberta and in Quebec, Labrador, and New Brunswick.

At least two species of *Zygadenus* (three, according to some botanists) occur in Canada: *Z. elegans* Pursh and *Z. venenosus* S. Wats. These are perennial herbs with mostly basal, grass-like leaves, growing from oval-shaped bulbs that are onion-like, but lack an onion odour. The flowers are relatively small, and cream-coloured to greenish white, borne in a compact to open cluster at the end of a slender, erect stem. These plants grow in damp, open meadows, prairies, and open woods. *Z. elegans* occurs from the Yukon and Northwest Territories southwards and eastwards to northern New Brunswick, and *Z. venenosus* from southern British Columbia to southwestern Saskatchewan. A third species, *Z. paniculatus* (Nutt.) S. Wats., is often considered to be a variety of *Z. venenosus*.

Indian hellebore and death camas have caused the deaths of livestock and humans; in many areas, death camas is the worst killer of sheep on the spring range of any poisonous plant. The symptoms of poisoning by these genera include a burning sensation of the mouth and throat, heavy salivation, nausea and gastrointestinal irritation, prostration, and, in some cases, hallucinations.

At some growth stages, some poisonous plants, especially water hemlock, resemble well-known edible species, and in the past even native people, supposedly familiar with the wild edible species, have died from mistakenly eating water hemlock. During the course of ethnobotanical research in the British Columbia interior, Nancy Turner was told of a number of instances of this having occurred within living memory of the Indian people consulted. In these cases, the species most often mistaken was water parsnip (*Sium suave* Walt.), whose roots and greens were commonly eaten, but which closely resembles water hemlock and grows in similar habitats. Because of the serious possibilities of confusing the two plants, we cannot recommend water parsnip as a wild green vegetable, and caution everyone against its use.

There is also a possibility of confusing death camas with wild onions because of the similarity of their bulbs (some even call death camas "poison onion"), see page 57. However, the bulbs and plants of death camas have no onion smell, so can be easily distinguished. They can be, and have been, confused with the well-known blue, or edible, camas (*Camassia* species) of western North America, the bulbs of which were a staple food of many Indian peoples. (Since blue camas is not a green vegetable, it is not included in this book.)

You should also remember that just because one part of a plant is recommended, it cannot be assumed that the entire plant, at any stage of growth, can be used. Usually the younger growth is milder and better tasting and contains less potentially toxic compounds than the mature growth, but this is not always the case. The opposite is true with water hemlock, where the newly expanding leaves are more poisonous than the mature leaves, according to Kingsbury.

Therefore, be careful to follow the instructions and recommendations in this and other reference books, and check the descriptions and illustrations thoroughly. If you have any doubts at all, leave the plant alone.

Finally, when using plants growing in or near water, be sure that you are not in an area polluted by domestic sewage or industrial wastes. If gathering from roadsides, edges of fields, or even forested regions, you should avoid areas that have been sprayed with insecticides, herbicides, or chemical fertilizers. Plants growing right next to roadways should be avoided because they may contain excessive lead from car exhaust. Thorough washing of the greens is always a good idea, especially if they are collected anywhere near human settlements.

Moderation and common sense are the keys to enjoying edible wild greens. The reader is advised to pay particular heed to the Warning notes included for some of the plant species that may be harmful or could be confused with poisonous species.

Gathering and Preparing Wild Green Vegetables

The collection and preparation of wild greens usually require little more than the use of a container and a small knife for harvesting and peeling the greens. For some species, such as cow parsnip, you should also wear gloves until the shoots are peeled, because the outer part and leaves may irritate your skin. Some people are very badly affected by this plant, and can get a rash similar to that caused by poison ivy by handling the unpeeled plants.

Most wild greens are at their best when young and tender, in the springtime before the plants flower. However, many are still quite acceptable when mature, especially if you harvest only the stem tips, where the youngest growth is found. With some plants, such as sea-rocket, you can harvest several successive crops from the same plants because as the tips are selectively removed, more will sprout to replace them.

In the interests of good conservation, you should be very selective in your harvesting activities. Ostrich-fern fiddleheads are particularly vulnerable to being over-harvested. The fiddleheads are the young unexpanded leaves of the plant, and cannot regenerate once they are removed. Therefore, if all the fiddleheads are taken from a plant it will die. However, if only one or two are removed from each of several plants, all will survive and serve as a source of greens year after year.

Some greens are best eaten raw, in sandwiches and salads. Others are somewhat strong-tasting and are improved with cooking. Some, such as the wild onions, should be treated merely as flavourings, since they are usually too strong in flavour to be used alone.

As with all cooking, attractive presentation of a wild greens dish is important in determining how it will be received. A pretty serving dish and a few whole green leaves or edible flowers as garnish can increase the appeal considerably. Choosing the best foods to complement the greens is also a consideration. Most of the salad dishes are good when accompanied by toast, muffins, or crackers. We have found that many of the seaside greens form a good combination with seafoods such as crab, prawns, oysters, salmon, or sardines. Serving other wild foods with the greens is also a good idea: what about a wild dinner of venison, partridge, wild rice dressing, wild greens and wild berries for a gourmet menu?

Like other green vegetables wild greens can be frozen, dehydrated, or even canned for future use. We recommend that you follow the processing directions for the cultivated greens most like the wild ones you want to preserve. For example, strawberry spinach can be processed like garden spinach, cow parsnip and "Indian celery" like true celery, and fiddleheads and other shoots like asparagus. The blanching time for freezing or the time required for drying will vary according to the size and thickness of the greens. Canning vegetables is more difficult, since improperly canned foods can cause food poisoning, and the processing temperatures and time required can reduce the nutritive content, but by following the usual canning methods, you should be able to can most wild greens successfully.

Format

The plants discussed in this book are listed in an order that is partly botanical and partly practical. The so-called lower plants—in this case, the marine algae, lichens, and ferns—are listed first, followed by the flowering plants, maintained within the broad categories of monocotyledons and dicotyledons. Within these two last sub-groups the species are listed alphabetically by scientific name (genus and species) within their families, which are also presented alphabetically. A total of twenty-six species or groups of related species in twenty different families (or divisions in the case of the lower plants) are included.

For each species or group of related species the most commonly applied colloquial names of both the plant and its family are included in the top outer margin. The corresponding scientific names are given at the bottom of the page. When two or more related species can be used similarly, they are treated in the same section. Alternative or localized common names are given in the text.

The use of scientific names is necessary to indicate exactly which species is being discussed. Common names can be ambiguous because they may apply simultaneously to a group of related species, or to two or more completely unrelated species. Furthermore, in different localities different common names may be used, and, of course, the common names vary with the different languages spoken in Canada and around the world. Hence, scientific names become, at least ideally, a universal identification label for each and every plant species.

Unfortunately, the problem of naming plants will never be completely solved, even with the use of scientific names, because botanists still may disagree about which scientific name is valid and whether a taxonomic grouping constitutes a species or merely a subspecies or variety. As more information becomes known about a plant, its status and classification may change. This explains why there is the occasional synonym even with a scientific name. We have included such synonyms in cases where we felt their omission might cause confusion. In general, we have followed the scientific nomenclature used in our major botanical reference works: *Vascular Plants of the Pacific Northwest* by C. Leo Hitchcock *et al.*; *Manual of Botany* by Asa Gray; *The New Britton and Brown Illustrated Flora of Northeastern United States and Adjacent Canada* edited by H. A. Gleason; *Illustrated Flora of the Canadian Arctic Archipelago* by A. E. Porsild; and *Flora of Canada* by H. J. Scoggan.

Since this book is written mainly for non-botanists, we have tried to minimize the use of technical botanical terminology. However, because it was not always possible to avoid unfamiliar terms, we have provided a glossary at the end of the text.

The information contained in this book was derived in part from personal experience but also from a wide variety of reference books and articles, all of which the reader will find listed in the bibliography.

The following table includes the weedy species already discussed in *Edible Garden Weeds of Canada,* Edible Wild Plants of Canada Series, No. 1.

Wild Green Vegetables Discussed in
Edible Garden Weeds of Canada

Common Name	**Scientific Name**	**Edible Green Part**	**Preparation Required**
Green amaranth	*Amaranthus retroflexus* L.	leaves, shoots	raw in salads or cooked like spinach
Common burdock	*Arctium minus* (Hill) Bernh.	young leaves, stalks	raw or cooked like asparagus (stalks must be peeled)
Milkweeds	*Asclepias syriaca* L., *A. speciosa* Torr. and relatives	young leaves, shoots, half-ripe pods	always cooked; used like asparagus
Orache	*Atriplex* species	leaves, young stems	raw in salads or cooked like spinach
Wild mustards	*Brassica* species	young basal leaves	raw in salads or sandwiches or cooked like cabbage
Shepherd's purse	*Capsella bursa-pastoris* (L.) Medic.	young basal leaves seeds, pods	raw in salads or sandwiches or cooked like cabbage; used as seasoning
Lamb's-quarters	*Chenopodium album* L. and relatives	leaves, young stems	raw in salads or cooked in same way as spinach
Chicory	*Cichorium intybus* L.	young basal leaves	raw in salads or cooked like dandelion greens

Common Name	Scientific Name	Edible Green Part	Preparation Required
Thistles	*Cirsium arvense* (L.) Scop. and *C. vulgare* (Savi) Airy-Shaw	young leaves, stalks	raw in salads or cooked; (prickles must be trimmed off, stalks peeled)
Stork's-bill	*Erodium cicutarium* (L.) L'Hér.	young leaves	raw in salads or cooked
Bedstraw	*Galium aparine* L.	young leaves, shoots	cooked (too rough to eat raw)
Wild lettuce	*Lactuca* species	young leaves	raw in salads or cooked like dandelion greens
Nipplewort	*Lapsana communis* L.	young leaves	raw in salads or cooked like dandelion greens
Common peppergrass	*Lepidium densiflorum* Schrad.	young leaves	raw in salads or sandwiches, or cooked in same way as cabbage
Common mallow	*Malva neglecta* Wallr.	young leaves, young seedpods	raw in salads or cooked—especially good in soups
Miner's-lettuce	*Montia perfoliata* (Donn) Howell	leaves, stems	raw in salads or cooked like spinach
Common evening-primrose	*Oenothera biennis* L.	young leaves	cooked like dandelion greens

Common Name	Scientific Name	Edible Green Part	Preparation Required
Yellow wood-sorrels	*Oxalis* species	young leaves, stems	raw in salads, as flavouring, or cooked like sorrel; good in sauces
Common plantain	*Plantago major* L.	young leaves	raw in salads or cooked like dandelion greens
Giant knotweeds	*Polygonum cuspidatum* Sieb. & Zucc. and *P. sachalinense* F. Schmidt *ex* Maxim.	young shoots	raw in salads or cooked like rhubarb stalks
Purslane	*Portulaca oleracea* L.	leaves, shoots	raw in salads or cooked like spinach
Watercress	*Rorippa nasturtium-aquaticum* (L.) Schinz & Thell.	leaves, shoots	raw in salads or sandwiches, or cooked like cabbage
Sheep sorrel	*Rumex acetosella* L.	young leaves	raw in salads as flavouring or cooked like garden sorrel; good in sauces
Dock	*Rumex crispus* L. and relatives	young leaves, shoots	cooked like spinach
Common hedge mustard	*Sisymbrium officinale* (L.) Scop.	young basal leaves	raw in salads or cooked like cabbage

Common Name	Scientific Name	Edible Green Part	Preparation Required
Sow-thistles	*Sonchus arvensis* L., *S. asper* (L.) Hill, and *S. oleraceus* L.	young leaves	raw in salads or cooked like dandelion greens; prickles must be trimmed off
Chickweed	*Stellaria media* (L.) Cyrill.	young leaves, shoots	raw in salads or cooked like spinach
Comfrey	*Symphytum officinale* L.	young leaves	cooked like spinach
Common dandelion	*Taraxacum officinale* L.	young leaves, flower buds	raw in salads or cooked, preferably in two changes of water
Pennycress	*Thlaspi arvense* L.	young leaves	raw or cooked, as flavouring
Salsify and Goat's-beards	*Tragopogon porrifolius* L., *T. dubius* Scop. and *T. pratensis* L.	young leaves	cooked with their own roots
Red clover and white clover	*Trifolium pratense* L. and *T. repens* L.	young leaves	raw in salads or cooked (use sparingly); used as tea
Stinging nettle	*Urtica dioica* L.	young leaves, shoots	cooked like spinach; do not eat raw

Wild Green Vegetables

Dulse, Laver, and Other Sea Vegetables

(Seaweeds)

Palmaria palmata (L.) Kuntze, Porphyra species, and other marine algae

How to Recognize

There are dozens of edible seaweeds to be found along Canadian shorelines. In fact it is probable that, except for one genus, most marine algae are edible, though some might not be palatable. The one exception is the genus, *Lyngbya*, in which some members are known to be poisonous even in small amounts (see Warning), but their unappetizing appearance is unlikely to attract the forager. Algae are among the most nutritious plants on earth and, although recognized as such in other parts of the world, they are little appreciated by most Canadians, mainly because they are an unfamiliar food.

Dulse and laver are both red algae, in the phylum Rhodophyta. The well-known dulse [*Palmaria palmata*, previously called *Rhodymenia palmata* (L.) Grenville], already widely used as a vegetable, is rose-red to reddish purple, up to about 40 cm tall, growing from a tiny disc-shaped holdfast. The blades are flattened, thin and lobed, making the plant appear hand-shaped. The older blades often have rows of smaller leaflets attached to the margins.

There are many species of laver in the genus *Porphyra*. (In this genus is the famous *nori* of the Japanese.) The lavers are simple, membranous plants, somewhat resembling thin sheets or wide strips of wet, transparent rubber. The edges are smooth or irregularly lobed, depending on the species and growth stage. In colour the lavers vary from light pink to dark reddish purple or brownish purple when fresh, appearing considerably darker, almost black, when dry. Some species, such as the common purple laver (*P. perforata* J. Agardh), can grow up to 1.5 m or more in length, but the plants are best for eating when small, young, and tender. The blades are stalkless and are attached to rocks by a small holdfast.

Another well-known, edible red alga is Irish moss, or carrageen, (*Chondrus crispus* Stackhouse). It is a dark-red to greenish or whitish seaweed growing in dense, erect clumps up to 10 cm tall. Several flattened blades rise from a single holdfast, branching repeatedly into a series of narrow to rounded segments, giving the plants a fan-like, or bushy, appearance.

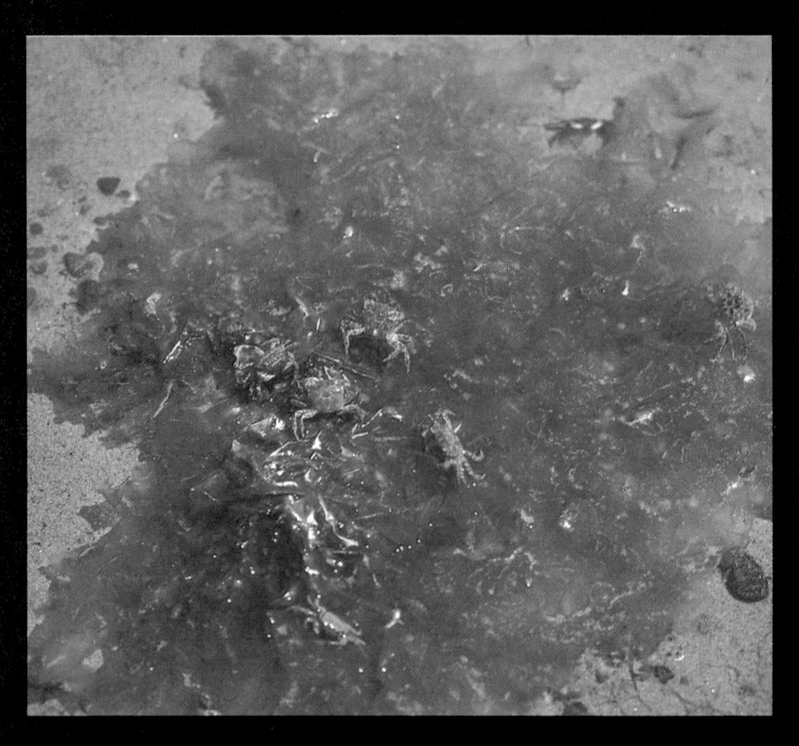

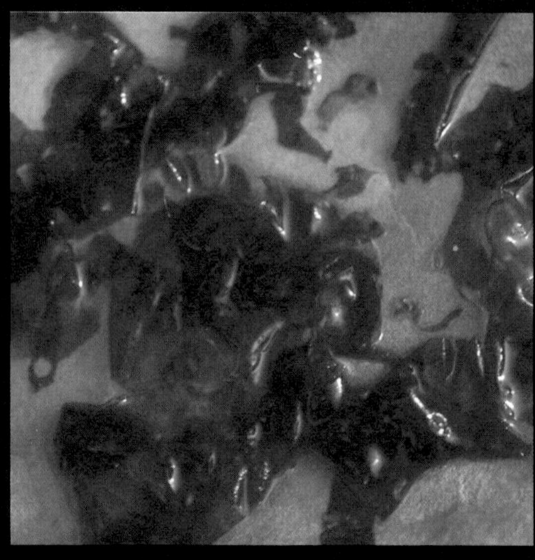
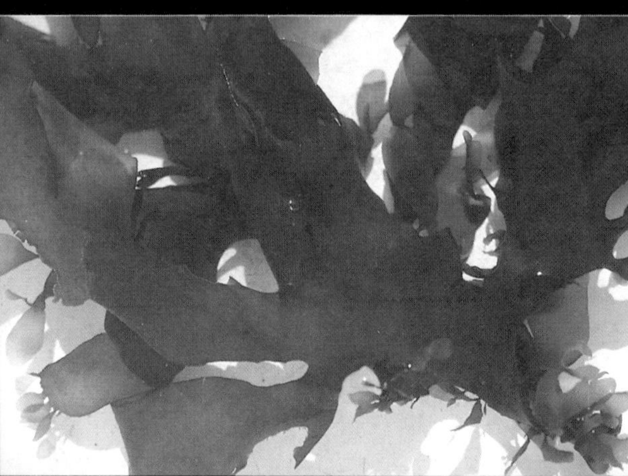

Many of the kelps, brown algae in the phylum Phaeophyta, also make delicious vegetables. Prominent among these are bull kelp, or bull-whip kelp [*Nereocystis luetkeana* (Mertens) Postels & Ruprecht], giant kelp [*Macrocystis pyrifera* (L.) C. Agardh, *M. integrifolia* Bory], and various species in the genera *Laminaria* and *Alaria*. Bull kelp, well known on the Pacific coast, is olive-green and consists of a long, hollow, whip-like stipe terminated by a large bulbous float to which are attached many long, thin blades. Mature plants may exceed 25 m in length. The holdfast is fleshy and root-like. Giant kelps are pale yellow-brown with long, slender stipes up to 25 m or more in length. Large, flattened, sickle-shaped blades rise at intervals along the stipe. The surface of the blades is bumpy or wavy and each is subtended by a small oval float at the point of attachment to the stipe. Several stipes may arise from one holdfast.

The laminarias and alarias, often referred to as short kelps, are quite variable, but most are olive-tan to dark brown and have a leathery texture. The stipes are thick, tough, and solid, usually quite short, and the blades are flat and smooth or wavy edged. In some species they are simple, long, and tapering, in others deeply cut into segments. The blades of *Laminaria* species do not have a midrib, although sometimes the median section is smoother than the edges. In *Alaria* species there is a narrow midrib, and smaller fertile blades arise along the stipe at the base of the main blade.

Among the edible green algae, in the phylum Chlorophyta, two prominent species should be mentioned: sea lettuce (*Ulva lactuca* L.) and tube seaweed, or green nori [*Enteromorpha intestinalis* (L.) Link]. Both are bright-green, membranous, transparent species. The blades of sea lettuce are wide and irregularly shaped, and those of tube seaweed are long, tubular, and inflated with gas.

For a description and discussion of numerous other edible seaweeds to be found on North American coasts, the reader is referred to a detailed and informative volume, *The Seavegetable Book* by Judith Cooper Madlener.

Ulva lactuca
Palmaria palmata

Nereocystis luetkeana
Porphyra perforata

Where to Find

Most of the sea vegetables mentioned here are perennials, but are at their peak of growth from early spring to fall. Some of the lavers and the bull-whip kelp are annuals. Dulse and various laver species are found in both Atlantic and Pacific waters, as are the laminarias and alarias, sea lettuce, and tube seaweed. Irish moss is found only on the Atlantic coast, and bull and giant kelps only along the Pacific coast. Most grow on rocks from the mid-tide mark to below the low-tide mark. Bull and giant kelps grow in extensive beds in the subtidal zone. Sea lettuce and tube seaweed tend to grow in rocks and pools of the upper tidal zone, and often occur in brackish water.

How to Use

Edible seaweeds are at their prime when young, fresh, and tender. This means gathering them while they are still growing or immediately after a storm, when they have been dislodged and washed up on the beach. If the ocean water is clear and clean, you can simply wash off your harvest at the site, being careful to remove any sand and grit adhering to the plants. You can wash the seaweeds in fresh water, but be careful not to soak them too long, as some of the nutrients may be lost. Immersion for less than a minute in lukewarm water will not be detrimental, but do not use hot water, as it could cause your sea vegetables to deteriorate in texture, taste, and nutritive value.

Seaweeds can be eaten fresh and raw in salads, cooked in soups, stews, and casseroles, or stir-fried Oriental fashion. The best way to preserve them for later use is by sun-drying them. Spread them out in thin layers on rocks, clean grass, towelling, or fibreglass screening, or hang them over a clothesline or drying rack. Turn them frequently, every hour or so, until they are crisp and completely dried. The dried seaweeds should then be stored in an air-tight container in a cool, well-ventilated place. If they should pick up a little moisture during storage simply place them in the oven at about 40°C (100°F) for a few minutes to restore their crispness. Dried seaweeds can be eaten as is, as a nibble, or can be rehydrated. Soak them from 1 to 2 hours in fresh water—enough barely to cover the seaweed—and, to preserve all the nutrients, use this same water in cooking or in making accompanying sauces or gravies, unless it is too strong-tasting. Many seaweeds, except the kelps, can be successfully frozen. Kelps, however, can be pickled in vinegar and make a tasty relish.

You can tenderize the tougher species of *Porphyra* and *Laminaria* by marinating them in vinegar or soy sauce, or by shredding and leaving them in a dish or plastic bag for a day or so before use.

Sea vegetables produce natural gels that can be used in place of animal gels (derived from the hoofs and cartilage of cattle and other animals) to make a great variety of puddings and gelatin desserts. Algal gels are already being widely used commercially as stabilizers and gelatins in many prepared foods, such as yoghurt, ice cream, candies, breads, and coatings for canned meat and fish. Irish moss, which is particularly good for making gelatin desserts, contains high concentrations of the gel carrageenan, produced by many of the red algae. Agar is a well-known gel derived from some red algae and is sold commercially in powdered, stick, or flake form. Both carrageenan and agar are water-soluble and can be easily extracted from algae by boiling the plant in water or milk. Unlike animal gels, agar will set at room temperature in a few minutes. The kelps and other brown algae produce another widely used food gel known as algin.

As a group, the marine algae are extremely rich in nutrients. Judith Madlener in *The Seavegetable Book* (p. 24) states that, "The nutritive values of many sea vegetables greatly exceed those found in any existing food source". Of course, the concentration of different nutrients varies considerably depending on the species and type of alga, but, in general, seaweeds are high in vitamins, digestible proteins, carbohydrates (in the form of gels), fats and oils, minerals, and essential trace elements. In a detailed discussion of sea vegetables and nutrition Judith Madlener claims that dulse contains, by weight, half the vitamin C found in oranges; lavers, sea lettuce, and some brown algae contain even more. (However, samples of laver and dulse measured by Hoffman *et al*. in "Ascorbic Acid and Carotene Values of Native Eastern Arctic Plants" were found to contain no vitamin C in the case of laver and only 10.0 mg per 100 g fresh weight for dulse. There could be several reasons for the discrepancy in these findings: the samples may have been collected at different times or seasons; different analytical procedures might have been used; or there might have been ecotypic variations in the laver and dulse used as samples for analyses.) Some laver species have been found to contain up to five times more vitamin A than is found in chicken liver.

The proportion of protein in algae can be as high as 25 per cent of the dry weight, and on the whole, algal proteins are reasonably easy to digest, and nutritionally similar to soy protein. One hundred grams of dried laver (Japanese *nori*) can supply one-half the daily adult protein requirement. It has recently been found that algal carbohydrates, previously thought to be indigestable by humans, are capable of being digested when present in the diet for about a week; the body can, in this time, develop new enzymes and intestinal organisms to deal with the new food. Fat content of sea vegetables can be as much as 4 per cent of the dry weight (usually 1 per cent or less). Some algae contain more calcium by weight than whole milk, and others contain as much iron as in whole wheat. Given these facts, it is amazing that more people do not take advantage of these maritime foods, free for the taking along the coastlines of Canada.

Even those people living away from the coast can enjoy sea vegetables since many seaweeds and seaweed products can be purchased from health-food shops and from Japanese and other ethnic markets in all major cities and towns in North America. Dried dulse is gathered commercially in the Maritimes and can be purchased in many groceries and supermarkets.

Warning

Do not gather seaweeds near centres of population, where they may be contaminated by sewage or industrial wastes. The only toxic macroscopic algae in themselves are certain members of the genus *Lyngbya*, in the phylum Cyanophyta. They are dark bluish-green, filamentous algae, each filament smaller than the diameter of a human hair. They form dense, irregular floating mats or cling to submerged vegetation, are common in salt marshes, and are found on both Atlantic and Pacific coasts.

Members of a genus of brown algae, *Desmarestia*, contain esters of sulphuric acid and could cause digestive upset, but the sour taste makes it unlikely that they would be eaten in any quantity.

Although seaweeds are rich in nutrients we recommend caution in using this food, because of its high iodine content. Less than 0.2 g dried kelp per day will satisfy an adult's daily iodine requirement. Iodine requirements in man are paradoxical: both a deficiency as well as an excess can cause goitre. With the use of iodized salt and iodophors on dairy farms, the dietary daily intake of iodine in North America now actually provides more than the recommended daily intake, especially in children.

We therefore do not recommend the regular daily supplementation of the diet with iodine from seaweed. The occasional consumption of kelp or other seaweeds should not, however, present a hazard since iodine is rapidly excreted from the body.

Suggested Recipes

Laver Soup*

10 mL	sesame seed oil	2 tsp
	1 small onion, thinly sliced	
	4 strips dried laver, about 5 by 15 cm, crumbled	
250 mL	snow pea pods	1 cup
	1 thin slice fresh ginger root, minced	
15 mL	soy sauce	1 tbsp
500 mL	fish stock, heated	2 cups

In a saucepan, heat oil and sauté onion until it becomes translucent. Stir in remaining ingredients and bring to a boil. Reduce heat. Simmer for 5 minutes, covered. Serve hot. Serves 4.

*All these recipes are adapted from Judith Cooper Madlener's in *The Seavegetable Book*.

Stuffed Laver Fronds*

250 mL	pearl barley	1 cup
750 mL	beef broth	3 cups
250 mL	chopped onion	1 cup
	2 cloves garlic, chopped	
45 mL	butter *or* margarine	3 tbsp
	sea salt and freshly ground black pepper	
	6 large laver fronds (fresh *or* dried)	
	corn oil	

Place the barley in a heavy skillet and dry roast for a few minutes until it turns a light-yellow colour and begins to smell nut-like. Place in a deep, covered pot. Add broth, onions, garlic, butter *or* margarine, salt, and pepper. Simmer gently, stirring occasionally, for 1 hour or until all the broth is absorbed. Dip dried laver fronds in boiling water for 1 second to make them pliable. Drain and spread out on waxed paper. (This step is not necessary if fresh fronds are used). Place a

suitable amount (depending on size of frond) of the barley stuffing in each frond and roll up from narrow end. Sauté the stuffed rolls in a little corn oil for about 5 minutes until the laver is tender. This dish is a good choice for a large dinner party since most of the preparation can be done ahead of time. Serves 4–6.
Note: as a variation, sauté 10 mL (2 tsp) ground cumin for a minute or two in a frying pan, then add about 125 mL (1/2 cup) ground beef. Sauté until meat is cooked and add 10 mL (2 tsp) tumeric. Add to cooked barley mixture, and stuff laver fronds as before.

Baked Dulse Salmon Loaf*

30 mL	dried dulse	2 tbsp
125 mL	whole milk	1/2 cup
175 mL	cracker crumbs	3/4 cup
	3 eggs, unbeaten	
	juice of 1/2 lemon	
15 mL	parsley, chopped	1 tbsp
5 mL	sea salt	1 tsp
	1 large can salmon	
	1 lemon, cut in wedges	

Chop dried dulse finely. Warm milk and mix in cracker crumbs, then all other ingredients. Preheat oven to 180°C (350°F). Lightly grease the inside of a mould. (A flavourless oil such as corn oil is best.) Fill the mould with salmon mixture and bake 1 hour. Unmould and serve hot or cold with lemon wedges. Serves 4.

Irish Moss Blender Pudding*

175 mL	dried Irish moss	3/4 cup
1 L	whole milk	4 cups
500 mL	fresh strawberries, puréed	2 cups
125 mL	honey	1/2 cup
	pinch of sea salt	

Soak the Irish moss for 30 minutes in enough cold water to cover. Wash well. Drain and pick over to remove any foreign matter. Pour milk into the top of a double boiler. Place the Irish moss in a square of cheesecloth, tie up the ends, and suspend the bag in the milk. Simmer for 30 minutes. Press the bag against the side of the pan occasionally to release the gel. Stir frequently. Remove from heat, discard the spent bag, and pour the milk mixture into the blender. Add the puréed strawberries, honey, and salt, blend at high speed, and pour into dessert dishes. Cover tightly and refrigerate for several hours before serving. Serves 6.

Fried Sweet Laminaria Chips*

Cut dried laminaria fronds into 3 by 5 cm pieces. Heat a small amount of oil in a heavy skillet and sauté the kelp pieces a few at a time, adding more oil as necessary. Drain on absorbent paper towelling. Sprinkle with raw sugar and serve as a snack or appetizer.

Bull Kelp Sweet Pickles*

5 L	bull kelp stipes, sliced	4 qt
1 L	cider vinegar	4 cups
500 mL	water	2 cups
1.5 kg	raw sugar *or* honey	3 lb
	3 large onions, sliced	
	8 slices lemon	
	6 sticks cinnamon	
30 mL	whole cloves	2 tbsp
15 mL	mace	1 tbsp
	1 pimiento, sliced	

Pare off the outer skin of the kelp stipes with a potato peeler. (One large bull kelp stipe will make approximately 500 mL or 2 cups of pickles.) Rinse the peeled stipes in cold water and cut into slices about 5 mm in thickness. Soak the cut pieces in fresh water for three days, changing the water several times to remove bitter sea salts. On the fourth day, place the kelp in a saucepan and cover with cold water. Bring to a boil and simmer 12 to 14 minutes. Combine the vinegar, water, sugar *or* honey, onions, lemon slices, spices, and pimiento. Bring to a boil and simmer 5 to 10 minutes. Pour over the drained rings and let stand overnight. Next day, drain off syrup, heat it to boiling point and pour it over the rings again. Let stand another night. On the sixth day, drain off the syrup and bring it to a boil again. Fill hot, sterilized jars with the kelp rings, cover with the hot syrup, and seal. Let stand at least a month before serving. Makes about 20 medium jars.

Sea Lettuce Salad*

1 L	fresh sea lettuce fronds	4 cups
125 mL	fresh cream	1/2 cup
10 mL	fresh lemon juice	2 tsp
10 mL	cider vinegar	2 tsp
15 mL	olive oil	1 tbsp
	pinch of cayenne pepper	
125 mL	onions, sliced into thin rounds	1/2 cup

Wash the sea lettuce quickly in lukewarm water. Pat dry with a clean towel. Chop finely. Prepare dressing by blending together cream, lemon juice, vinegar, oil, and cayenne pepper. Pour over sea lettuce and toss with onion rounds. Serves 2–4.

More for Your Interest

Seaweeds are valued as food in many parts of the world, especially in Japan, Korea, China, Polynesia, and the Philippines, as well as in the West Indies, Chile and Peru, New England, the British Isles, and other temperate North Atlantic countries. In South Wales, for example, a traditional breakfast food consists of laver coated with oatmeal and fried in hot bacon fat. This dish, called laverbread, is served in most fish and chip shops in Wales and southern England.

In Canada, dulse is especially popular in the Maritimes. It is harvested there on a commercial scale, partially dried, and pressed into plugs (cakes) or sold in sheet-form in markets throughout the country. Some eastern pubs offer dulse chips with beer in the same way as potato chips or peanuts are served elsewhere. Irish moss is also harvested commercially in the Maritimes.

In British Columbia, laver (*Porphyra perforata*) has long been a favourite food of the coastal Indians and is still enjoyed today by them and by the immigrant Chinese and Japanese populations. It is gathered in the spring, shredded, and dried on rocks in the sun. In the old days it was often left in a large pile to ferment for four or five days to tenderize it before being dried. It was eaten with grease rendered from the eulachon fish, or with boiled fish or clams, or was simply nibbled in dried form. A modern preparation is made by boiling it with canned creamed corn. The coastal Indians also ate giant kelp and some of the laminarias when covered with herring spawn. They still gather the spawned-on seaweeds today and fry them in bacon grease or steam them. Bull kelp was not used as food by the Indians, but was an important material for making fishing lines and anchor ropes, and the hollow bulbs were used to store fish oil and other liquids.

Rock Tripe and Other Edible Lichens

(Lichens)

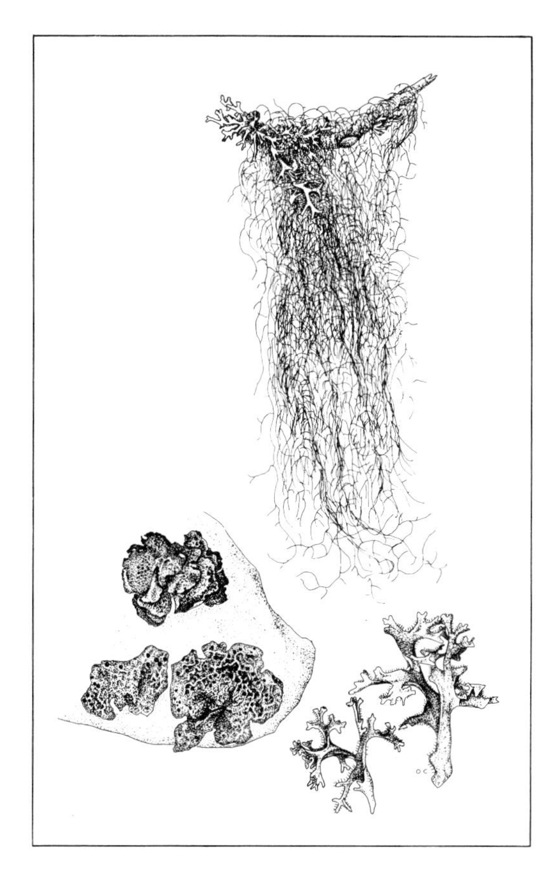

Other Name
Rock tripe is also known by its French name *tripe de roche*.

How to Recognize
Many lichen species have been used as food by man in different parts of the world. In Canada, rock tripe is undoubtedly the best-known edible lichen and certainly one of the most important as a survival food. The *Umbilicaria* species, several of which are found in Canada, are foliose lichens with flat, leaf-like thalli often 5 cm or more across, attached to the rock substrate at a single central point. When dry, the thallus is rather brittle and greyish to deep brown in colour. When damp, it becomes somewhat limp and rubbery and is blackish or dark green. The underside of the thallus is usually darker in colour and often velvety or hairy. In some species the margins of the thallus are smooth; in others they are irregularly lobed or deeply cut. Rock tripe is usually sterile. The fruiting bodies, or apothecia, when they do occur, are small and black and are scattered on the upper surface of the thallus. Some members of the genus *Umbilicaria* were formerly considered to be in a separate genus, *Gyrophora*.

Umbilicaria   *Bryoria fremontii* *Cetraria islandica*

Umbilicaria species and other lichen species

Another lichen, little known as a food, is the black tree lichen, or tree hair [*Bryoria fremontii* (Tuck.) Brodo & D. Hawksw.—formerly known as *Alectoria fremontii* Tuck., in the "*Alectoria jubata*" complex]. This is a dark-coloured, filamentous species that hangs from the branches of trees, closely resembling a thick tuft of black hair. In the main part of its range, it may grow 25 cm or more in length. The texture of the thallus when dry is wiry; when damp it is soft and limp. This lichen sometimes produces tiny cushions of bright-yellow granules (soredia) or yellow, disk-shaped apothecia scattered along the filaments. There are a number of related but inedible species that closely resemble this one. These species often contain large quantities of certain organic substances (lichen acids) that render them bitter. One of these related species most closely resembling *B. fremontii* is called *Bryoria tortuosa* (Merr.) Brodo & D. Hawksw. and sometimes contains high concentrations of the potentially poisonous yellow compound, vulpinic acid. This species is distinguished by its twisted, tortuous filaments and long, slender, yellow stripes spiralling down the branches. Perhaps the best way to distinguish it is by its very bitter or sharp taste. *Bryoria fremontii* contains vulpinic acid only in the yellow granules, or fruiting bodies.

A third edible lichen, well known in northern Europe, is Iceland moss (*Cetraria islandica* Ach.). Growing in extensive patches on the ground, it has upright, flattened branches of varying shades of brown that grow up to 7 cm long and 1 cm wide. The margins range from slightly incurved to strongly curled, and bear a fairly regular fringe of short black spine-like appendages. When dry, the thallus is brittle, and when wet it is quite soft and pliable. Brown disk-shaped fruiting bodies sometimes are produced along the margins of the branches. Some related species also are edible.

Where to Find
Rock tripe grows mainly on exposed granite rock in upland areas, and various species are very abundant throughout northern Canada. Black tree lichen grows on branches of coniferous trees such as larch, pine, and Douglas-fir, in the montane forests of western Alberta and British Columbia. Iceland moss grows on upland moors and bare, stony soil especially in mountainous areas and in the Arctic tundra zone of North America and European countries.

How to Use
It is unfortunate that, of the hundreds of lichen species that occur in Canada, only a few are suitable for use as food. Most lichens are rendered inedible and very bitter by lichen acids, and must be leached in water or treated chemically with carbonates of sodium or of potassium before they can be used as food. Even the more edible species described here must be soaked for several hours in water or in a weak solution of baking soda before they are fit to eat. Otherwise they may cause digestive upset. Once the bitterness is removed, these lichens are palatable. They contain a variety of carbohydrates, of which polysaccharides are the most common. Unfortunately, it appears from nutritional studies, such as those cited by Llano that lichen carbohydrates are only partially digestible in humans. Since protein and fat content in lichens is very low, their main food value appears to be in providing bulk and filling the stomach in times of food scarcity. Lichens also provide some vitamins and minerals, but in most cases not enough to be nutritionally significant. The fact remains that there are many instances on record of lichens being used successfully to prevent starvation, and other cases where lichens have been used as a regular part of the human diet or even as a delicacy.

Suggested Recipes

Rock Tripe

Gather rock tripe by plucking or scraping it from the rock with a knife. Wash and cut off any gritty parts. Soak for several hours in two changes of water in which about 5 mL (1 tsp) baking soda per litre (quart) has been dissolved. Then drain, cover with water, and simmer slowly for an hour or so, until tender. The tripe becomes jelly-like in consistency and can be used as a thickener for soups and stews, and in an emergency can be eaten on its own after leaching. It can also be dried and powdered to make an acceptable "stretcher" for wheat flour, or it can be cooked with milk, eggs, and sugar to make a custard.

Black Tree Lichen, Indian Style

Gather lichen from tree branches, using a long pole to twist it off the higher branches, then clean it of any twigs and debris. Soak overnight in water, preferably running water such as a stream or river. Dig a pit about 1 m square and .75 m deep, line it with large, round rocks, and light a hot fire in it, allowing it to burn until the rocks are nearly red-hot. Remove the ashes, then place a layer of boughs or damp moss over the rocks. Pack the damp lichen on top in a layer 15 to 25 cm thick. Other food, such as wild onions, may also be added. Cover with more boughs or moss, then fill the pit with dirt, leaving a stick standing vertically at the centre. When the pit is full, remove the stick and pour water through the hole left, until you hear the rocks at the bottom hissing and cracking. Then seal the hole and leave the pit for about 24 hours. When the pit is uncovered, the lichen will have congealed into a jelly-like mass 2 to 3 cm thick. This can be cut up and eaten fresh, or sun-dried into cakes to be used later. The dried cakes should be soaked

in water to soften them before use. They are good in soups and stews and can be mixed with other foods such as saskatoon berries, or cooked with apples, raisins, molasses, or brown sugar.
Note: this lichen was an important food of the Interior Salish Indians of British Columbia, who prepared it as above. If you are not able to cook it in the traditional way, you can try boiling it in water for several hours in a heavy saucepan, or, to speed up the process, in a pressure cooker, but the native old-timers who remember the pit-cooked lichen will tell you that these modern methods produce an inferior product.

More for Your Interest
Many northern explorers and traders, including Sir John Franklin and Sir John Richardson, depended on rock tripe, or *tripe de roche* as it was called, for survival in times of famine. During Franklin's first expedition, rock tripe and scraps of leather were the only available foods for days at a time. Unfortunately the men were not able to leach the lichen properly and frequently suffered severe bowel complaints from their meagre diet. Rock tripe was also boiled and eaten by native peoples of the North, who cooked it with fish roe or other fish or meat.

Iceland moss is harvested commercially in Sweden, Norway, and Iceland. It is said that a new crop can be gathered from the same site every three years. The lichen is first cleaned, dried, and then reduced to powder, which is leached, like rock tripe, in water or a soda solution, then boiled to yield a jelly that forms the basis of various light, easily digested soups and puddings. The powder is also used in European countries to make porridge and bread. Considerable quantities of Iceland moss were formerly used in making ship's biscuits for sailors. The bread made with it was supposed to be less liable to infestations of weevils than when made from wheat flour alone.

Many lichens are important as animal fodder. One of the most significant is the reindeer lichen, *Cladina rangiferina* (L.) Nyl. (formerly *Cladonia rangiferina* Web.), a grey, shrubby species that, with related species of *Cladina*, grows in dense mats 10 to 15 cm high in forest and tundra regions of the Far North. It is one of the prime foods of reindeer in northern Europe and is also browsed by deer and cattle.

The black tree lichen, as well as being used as food by the Indians of British Columbia, was also utilized to make blankets, capes, and shoes, although it was not considered a prime material and was used mainly by poor people who had no animal skins for clothing. Iceland moss, being rather astringent, has been used in tanning leather, and both it and the black tree lichen have been fermented by distillers to produce alcohol.

Knowledge of the dyeing properties of lichens dates back to the days of the Old Testament and before. Lichens yield a wide range of dye colours and are still used today by many weavers. They are discussed in almost any book dealing with natural dyeing. Rock tripe yields a rich purple dye when treated with ammonia, fermented, and mixed with potash or soda. Iceland moss yields brown dyes of various shades, and black tree lichen produces a yellow dye. However, if you harvest these lichens for food or dyes, remember that they grow very slowly and should be gathered with discretion and only in places where they are very abundant.

Ostrich Fern

(Fern Family)

Matteuccia struthiopteris **(L.) Todaro**
(Polypodiaceae)

Other Names
Fiddlehead fern, edible fern.

How to Recognize
There are many ferns whose fronds have been eaten at the fiddlehead stage, when the young shoots are just unfurling in spring. Of these, ostrich fern is the best known, tastiest, and safest to eat. Others, such as bracken fern [*Pteridium aquilinum* (L.) Kuhn], cinnamon fern (*Osmunda cinnamomea* L.), and lady fern [*Athyrium filix-femina* (L.) Roth], are inclined to be bitter and unpalatable and are not recommended. Furthermore, there is recent evidence that bracken fiddleheads, though widely eaten, especially in Japan, are carcinogenic in many animal species, and are unsafe to eat in large doses (see Warning). No ferns should be eaten at their mature stage.

Ostrich fern is a perennial, growing in clumps from a deep, much-branched rhizome from which a stout, scaly stem-base rises vertically. The leaves, or fronds, are deciduous and of two types: sterile and fertile. The sterile fronds grow first, forming a ring of curled crosiers (fiddleheads) that gradually unfurl and expand into large, compound leaves. These have numerous long, narrow, coarsely toothed leaflets, or pinnae, spreading from both sides of the central axis in a feather-like formation. The fronds grow up to 1.5 m or more high and are broad, pointed, and feather-shaped in outline. After the sterile fronds appear, the fertile, spore-bearing fronds grow from the stem-base inside the sterile fronds. In contrast to the bright-green sterile blades, they are brownish to blackish, stiffly erect, and considerably shorter and more compact.

Bracken fern, described here because it has been widely used by wild-food enthusiasts but so that it can be avoided, is also a perennial, with fronds rising singly at intervals along a deep, black, branching, horizontal rhizome. The fronds are very large and deciduous, with tall (up to 2 m or more), stout, straw-coloured stems and coarse, divided blades that are triangular in outline. The spores are borne along the inrolled margins of some fronds, but reproduction is mainly vegetative, by means of the fast-growing rhizomes.

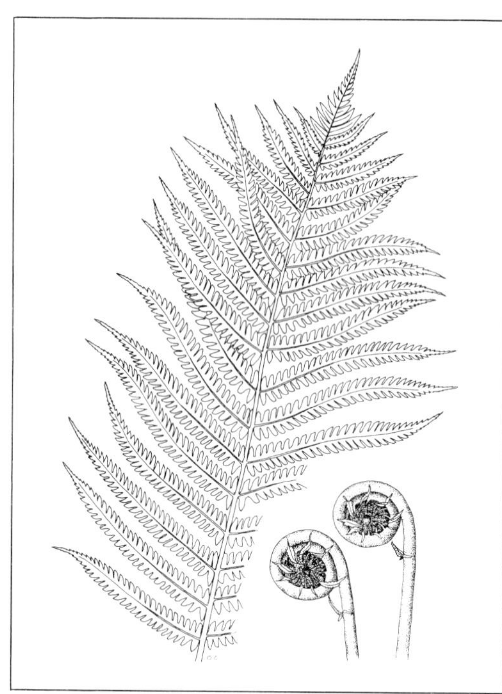

ostrich fern
(*Matteuccia struthiopteris*)

Where to Find
Ostrich fern grows in low, open ground and rich woods, particularly in alluvial soil, from Newfoundland to eastern and northern British Columbia and north as far as Great Slave Lake. It also occurs in the central and northern United States and in Eurasia. Bracken fern is one of the most widespread of all plants, growing in a wide variety of open habitats in almost all temperate and tropical regions of the world.

How to Use
The succulent, curled-up crosiers of ostrich fern can be broken off in spring and eaten raw as a nibble or cooked as a delicious green vegetable. The stouter stalks, when still tender and not more than 15 cm or so high, are broken off, washed, and cleaned of any scales. Please remember that these fiddle-heads eventually form the leaves of the plant and will not regenerate if picked, so be sure to leave plenty to enable the plant to survive. In some localities, entire populations of ostrich ferns have been devastated by eager fiddlehead harvesters. With judicious pick-ing, however, (as one would gather asparagus shoots) the same plants will yield fiddleheads year after year.

Some people enjoy fiddleheads raw, but in our opinion, they are far superior when steam-cooked until tender (about 10 minutes, depending on your own taste) and served with salt, pepper, butter, and lemon juice or sour cream if desired. The colour of the cooked shoots should be bright green; if they fade or turn yellowish, they are overdone.

Fiddleheads can be used in a variety of casseroles and other dishes and can be substituted for asparagus or green beans in any meal. They are easy to freeze, requiring only slight blanching first. Fresh ostrich-fern fiddleheads can be found as a specialty item in many eastern markets, and frozen fiddleheads are often available commercially in grocery stores and supermarkets throughout the country.

Fiddleheads are relatively high in iron. In a sample of 164 g of frozen, cooked fiddleheads there was found to be 1.3 mg iron, 363 mg potassium, and 25 mg vitamin C.

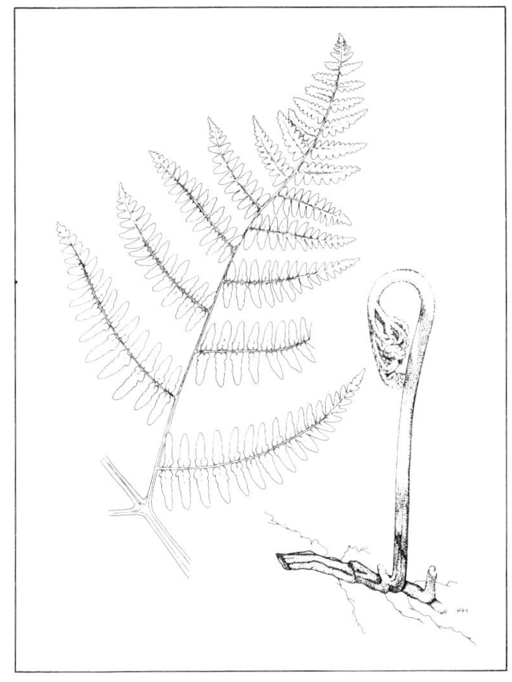

bracken fern (*Pteridium aquilinum*)
Caution: not recommended
see Warning

poison hemlock (*Conium maculatum*)
Poisonous: do not eat
see Warning

Warning
Never eat mature fronds of any fern species. It has been known for several years that bracken fern is toxic and carcinogenic to several animal species, but the active principle responsible for the carcinogenicity has not yet been identified. The young fronds as well as the mature plants are carcinogenic. Recently, feeding experiments at the Massachusetts Institute of Technology in which rats were fed either bracken fern or ostrich fern (collected in New Brunswick) confirmed that bracken was carcinogenic while ostrich fern at levels up to 10 per cent in the diet was not.

Bracken fern is widely consumed as a food in Japan, and it has been suggested that this might explain the high incidence of human stomach cancer there.

Note: see page 47 for description of bracken fern.

Be absolutely certain of your identification before sampling fern fiddleheads. A number of other plants, including the notorious poison hemlock (*Conium maculatum* L.), can be mistaken for ferns by the untrained observer. (Poison hemlock is actually a flowering plant in the celery family. It matures into a tall, branching, hollow-stemmed plant, but when young, its finely divided leaves are fern-like.)

Suggested Recipes

Steamed Fiddleheads

30 mL	butter *or* margarine	2 tbsp
500 mL	ostrich-fern fiddle-heads, washed	2 cups
30 mL	water	2 tbsp
	salt and pepper	

In a saucepan, melt butter *or* margarine, and heat until bubbly. Add fiddleheads and water, cover and cook at medium heat until stems are tender but still crisp and green (about 10 minutes). Season and serve hot. Serves 2–3.

Variations: add minced onion, Worcestershire sauce, and Tabasco sauce to taste; or serve with a dressing of fried, buttered breadcrumbs, mashed hard-boiled egg, and chopped parsley; or toss steamed fiddleheads in a dressing made with 50 mL (1/4 cup) lemon juice, 50 mL (1/4 cup) olive oil, 30 mL (2 tbsp) chopped green onion, a clove of garlic minced, chopped hard-boiled egg, parsley, salt, paprika, and sugar to taste.

Fiddleheads with White Wine

30 mL	butter *or* margarine	2 tbsp
500 mL	ostrich-fern fiddle-heads, washed	2 cups
30 mL	white wine	2 tbsp
30 mL	water	2 tbsp
	juice of 1/2 lemon	
50 mL	salted cashew nuts, coarsely chopped	1/4 cup
	a few powdered tarragon leaves	
	salt and pepper	

In a saucepan melt butter *or* margarine and heat until bubbly. Add fiddleheads, wine, and water, cover and cook at medium heat until fiddleheads are tender but still crisp (about 10 minutes). Add the lemon juice, cashew nuts, tarragon, salt, and pepper and serve hot. Serves 2–3.

Baked Fiddleheads

	2 dozen tender, young fiddleheads, washed, with tough ends removed	
500 mL	sliced unpeeled zucchini	2 cups
	1 medium onion, sliced	
	1 sweet red pepper, diced	
30 mL	butter *or* margarine	2 tbsp
	2 slices cooked, crumbled bacon	
	6 medium-sized mozzarella cheese slices	

In a saucepan, cover fiddleheads and zucchini with boiling water, simmer 5 minutes, then remove from heat and drain. Sauté onion and red pepper in butter *or* margarine until soft. Lay fiddleheads and zucchini slices in baking dish, sprinkle over onion, red pepper, and bacon, and cover with slices of mozzarella cheese. Bake at 150°C (300°F) for 20 minutes. Serve with toast as a light meal. Serves 2–3. (From Enid K. Lemon and Lissa Calvert, "Pick'n'Cook Notes".)

More for Your Interest

Bracken-fern rhizomes can be toxic when consumed raw and may have carcinogenic properties (see Warning above), but are high in nutritive value and were a major source of carbohydrate for many Indian groups, particularly on the Pacific Coast. The rhizomes were roasted over red-hot coals and pounded to remove the black outer skin. The whitish inner portion was eaten with oil or sometimes pounded into a kind of flour and formed into cakes. The tough central fibres were not eaten, but in some areas, such as on southern Vancouver Island, were saved, dried out, and used as tinder for starting fires, or placed between the two halves of a clamshell and ignited, where they would smoulder for hours. This "slow match" could be carried on journeys, or even buried at the campsite until nightfall when it would be used to light the evening fire.

Another fern whose rhizomes are edible, at least in small amounts, is licorice fern (*Polypodium glycyrrhiza* D.C. Eat.). This small fern grows on mossy rocks and tree trunks in western Canada. Though somewhat strong-tasting to the uninitiated, the rhizomes have a distinctive licorice flavour and at one time were considered as a commercial source of licorice. Indians used them as an appetizer and mouth-sweetener and also valued them as a medicine for colds and sore throats. For this last purpose they were simply chewed and the juice swallowed, or boiled in water to make a liquid cough syrup.

Wild Onions

(Lily Family)

Other Names
Wild garlic, wild leek, wild chives.

How to Recognize
The wild onions are herbaceous perennials growing from bulbs. Their most conspicuous feature is their unmistakable onion or garlic odour which becomes evident as soon as the plants are trampled or bruised. There are about ten species of wild onion in Canada, all of which are edible, although some are better tasting than others. Among the best species are: nodding onion (*Allium cernuum* Roth.); wild chives [*A. schoenoprasum* L. var. *sibiricum* (L.) Hartm.]; prairie onion (*A. textile* Nels. & Macbr.); wild garlic, or Canada onion (*A. canadense* L.); and wild leek (*A. tricoccum* Ait.).

Nodding onion is an attractive plant with elongated, pinkish-skinned bulbs, usually clustered. The leaves, several per bulb, are flattened, grass-like, and succulent, and persist during flowering. The flower stems, up to 50 cm tall, bear nodding terminal umbels of pink (rarely white) flowers. The heads turn upright after flowering. The fruits are 3-lobed capsules containing dull-black, flattened seeds.

Wild chives, forerunners of the garden chives, grow from thin, clustered bulbs. The leaves, usually only 2 per bulb, are hollow and cylindrical in cross-section. The stems are rather stout and up to 60 cm high. The flowers, pale to deep lilac, are clustered in compact, spherical heads. The petals and sepals persist after flowering, becoming papery when in fruit.

Prairie onion grows from ovoid, fibrous-coated bulbs. The leaves, 2 per bulb, are flattened and curved in cross-section. The flowering stems seldom exceed 25 cm in height. Each bears an upright umbel of white or pinkish-white flowers.

Wild garlic has ovoid bulbs, flattened, slightly keeled leaves, and stems up to 60 cm high. Unlike the other species, the terminal umbel of wild garlic bears few or no flowers, but consists chiefly of white or pinkish-white bulblets. These bulblets drop to the ground when mature to propagate new plants vegetatively.

(*Allium cernuum*)

***Allium* species**
(Liliaceae)

Wild leek grows from slender, oval, clustered bulbs up to 5 cm long, surrounding a short rhizome. The fleshy leaves are lance-shaped to elliptical, usually withering by flowering time. The slender flower stalk, up to 40 cm high, bears a many-flowered, hemispherical umbel. The flowers are white and the capsules deeply 3-lobed.

Where to Find

Wild onions occur throughout most of Canada, on dry hillsides, in open woods, and in damp meadows. Nodding onion is common on rocky bluffs and in open woods and meadows from British Columbia across the southern prairies and as far east as New York State in the United States. Wild chives grow in damp meadows and grassy slopes from northern and eastern British Columbia to Newfoundland, extending northwards into the Yukon and Northwest Territories. Prairie onion occurs on dry hillsides and plains from southern Alberta to Manitoba. Wild garlic is found in woods and meadows from western New Brunswick to southern Ontario.

Wild leek grows in rich woods and moist ground from southern Manitoba to Nova Scotia and New Brunswick. In some places it is becoming rare from over-harvesting and should be used only where it is very abundant.

wild onion (*Allium cernuum*)

How to Use

The different species of wild onions vary in quality and strength of flavour. Since they are usually stronger tasting than their domestic counterparts, use only half the amount called for in recipes where they are substituted for cultivated onions. As with domestic green onions, both the green leaves and the bulbs can be used. The young flower-bud heads also are edible. Wild onions are excellent as a flavouring in salads and sandwiches and are also delicious in soups, stews, casseroles, and in many other main course dishes. When well cooked they can be served alone as a vegetable, but many people find them too strong by themselves. We have provided a few suggestions on how to serve wild onions, but the possibilities are limitless.

One of Nancy Turner's favourite recollections to do with wild onions is of an outing of a university botany class to a shoreline along Vancouver Island's west coast. Edible mussels were abundant all along the rocky shore and a batch was soon harvested and set to bake dry in the coals of an open fire. As soon as the shells opened from the heat, revealing the bright-orange flesh of the mussel, a length of nodding onion was inserted into each. The mussels were left to simmer in their own juices a few minutes longer to impart the onion flavour and then were eaten straight from the shell. It would be hard to devise a more delectable feast.

death camas (*Zygadenus venenosus*)
Poisonous: do not eat
see Warning

Warning

Be careful not to confuse wild onions with the highly poisonous death camas (*Zygadenus venenosus* S. Wats.), also in the lily family. This species grows in meadows, open woods, and dry slopes from British Columbia to Saskatchewan. The bulbs and foliage have been responsible for the deaths of large numbers of livestock, especially sheep. The flowers are cream-coloured and borne in racemes, not umbels as in onions, and the plants and bulbs have no onion odour. White camas (*Z. elegans*), another poisonous *Zygadenus* species, is similar to death camas, but has broader leaves and open, loosely flowered clusters of greenish-white to yellowish-white flowers. It grows in damp, open meadows from the Yukon and Northwest Territories southwards into the United States and eastwards to northern New Brunswick. (See also p. 18.)

The consumption of large quantities of *Allium* species has been found to cause gastroenteritis in cattle and horses.

Suggested Recipes

Northern Omelette

30 mL	butter *or* margarine	2 tbsp
	6 large, fresh eggs	
30 mL	wild onion greens, finely chopped	2 tbsp
30 mL	young, peeled and diced, cow parsnip stems (see p. 79)	2 tbsp
	salt and pepper	
	a few wild onion blossoms for garnish	

Melt butter *or* margarine in skillet. Beat eggs, add greens, and pour mixture into skillet. Cook over medium heat until eggs set. Season with salt and pepper and serve immediately with onion blossoms as a garnish. Serves 3–4.

Wild Onion Flower-Bud Salad

	6 slices bacon, finely chopped	
50 mL	wild onion bulbs and greens, sliced	1/4 cup
125 mL	young wild onion flower buds	1/2 cup
30 mL	wine vinegar	2 tbsp
	2 celery stalks, diced	
	10 medium radishes, sliced	
	salt and pepper	

Dressing

250 mL	mayonnaise	1 cup
30 mL	chili sauce	2 tbsp
30 mL	green pepper, finely chopped	2 tbsp
10 mL	wine vinegar	2 tsp
5 mL	paprika	1 tsp
	3 hard-boiled eggs, finely chopped	

Fry bacon until crisp and brown, and drain off fat. Mix with wild onion bulbs, greens, and flower buds, vinegar, celery, radishes, salt, and pepper. To prepare dressing, mix together mayonnaise, chili sauce, green pepper, vinegar, paprika, and finely chopped eggs. Pour dressing over salad and mix lightly. Serve immediately. Serves 3.

Wild Onion Sauce

30 mL	butter	2 tbsp
30 mL	flour	2 tbsp
250 mL	scalded milk	1 cup
125 mL	wild onion bulbs and greens, chopped	1/2 cup
	salt and pepper	

Melt butter in heavy saucepan. Add flour and stir until smooth. Gradually add milk, stirring constantly. Add onion and seasoning and simmer 10 to 15 minutes over very low heat until sauce is thick and creamy. Serve over hot meat. Makes 375 mL (1 1/2 cups).

Wild Greens with Shrimp

750 mL	watercress shoots	3 cups
	10 wild onion bulbs and tops, thinly sliced	
	4 hard-boiled eggs, sliced	
500 mL	cooked shrimp, preferably fresh	2 cups
125 mL	cider vinegar	1/2 cup
75 mL	olive oil	5 tbsp
10 mL	sugar	2 tsp
	salt and pepper	
	2 or 3 medium mushrooms, thinly sliced (optional)	

In a large, flat salad bowl arrange in alternating layers: watercress, onion slices, egg slices, and shrimps. Mix together vinegar, olive oil, and sugar and sprinkle over salad. Season with salt and pepper and garnish with mushroom slices if desired. Serves 6.

More for Your Interest

Wild onions were a favourite food of some but not all Indian peoples in Canada. In the interior of British Columbia, nodding onions were gathered in large quantities and "barbecued" in deep underground pits until they were soft and sweet and had lost much of their sharp onion taste. They were often used to flavour other foods, such as black tree lichen (*Bryoria fremontii*), see page 44.

When Captain James Cook anchored at Nootka Sound on the west coast of Vancouver Island in 1778, his men noticed wild nodding onions, or "wild garlick", growing in profusion along the bluffs near the sea. When the Nootka Indians, who did not eat wild onions themselves, saw the Europeans digging and eating them, they realized that the onions would be a good item for trade. Thereafter when they visited Cook's ships, they brought wild onions along with animal skins and other articles to sell to the Europeans.

Wild onions, especially nodding onions, are easily culivated from seed or bulb. The plants are hardier and, in our opinion, more attractive than domestic onions and would certainly be an interesting addition to any garden.

Common Reed Grass

(Grass Family)

Other Names
Bennels, reed, reed grass.

How to Recognize
One of Canada's native grasses, this perennial plant often grows to 2 or 3 m in height. It spreads by means of fleshy, creeping rhizomes. The stems are stout and jointed, with hollow segments. Single, sheathing leaves with flat blades 20 to 40 cm long and up to 4 cm broad are borne at each node, at regular intervals along the stem. The flowering heads are terminal and plume-like, resembling those of the well-known pampas grass, but somewhat smaller. The heads are composed of numerous spikelets up to 1.5 cm long, each bearing 3 to 6 small flowers, almost obscured by dense clusters of long, silky hairs that give the head its white, feathery appearance. The fruits are small, but edible, grains. This grass is also known under the name *Phragmites australis* (Cav.) Trin.

Where to Find
Common reed grass grows around ponds, in marshes, ditches, and springs, and along riverbanks throughout most of southern Canada, except Newfoundland, and as far north as the District of Mackenzie. It is a cosmopolitan plant, being found in most temperate and many tropical regions of the world.

How to Use
The young shoots, just as they emerge from the ground or shallow water, can be dug up and eaten raw or prepared as one would asparagus or bamboo shoots. These shoots are best when still white and not exposed to the light. They are also said to make an excellent pickle. The partly unfolded leaves of more mature plants can be cooked as a potherb but become too tough by the time they are fully expanded. In addition, the starchy rootstocks can be dug in fall or spring and roasted or boiled as you would potatoes. The small grains do not separate easily from the hulls, but can be cooked with the hulls still attached, ground up, and eaten as a gruel. Fernald and Kinsey in *Edible Wild Plants of Eastern North America* say that they rate nutritionally between wheat and rice.

Warning
Be careful not to harvest the shoots, leaves, or rhizomes of this plant from areas where the water may be polluted by sewage, industrial waste, agricultural fertilizers, or pesticides.

***Phragmites communis* (L.) Trin.**
(Poaceae or Gramineae)

Suggested Recipes

Stir-Fried Reed Grass Shoots

500 mL	young, tender reed grass shoots, sliced diagonally	2 cups
15 mL	vegetable oil	1 tbsp
	1 clove of garlic, minced	
125 mL	green onions, chopped	1/2 cup
50 mL	cold water	1/4 cup
15 mL	soy sauce	1 tbsp
10 mL	corn starch	2 tsp
	dash of fresh *or* dried ginger, finely grated	
	salt and pepper	
	sesame seeds for garnish	

Wash reed grass shoots carefully, then drain, and set aside. In a skillet or Chinese wok heat oil over high heat until it starts to smoke. Quickly add garlic and stir for a few seconds until browned. Add reed grass shoots and onions. Blend together water, soy sauce, and corn starch and pour over shoots, stirring continuously. Add ginger and seasoning and continue stirring until shoots and onions are tender but not soggy. Garnish with sesame seeds and serve hot with rice. Serves 2.

Note: this basic recipe is open to many variations. Try replacing part of the reed grass shoots with thinly sliced celery, fresh mushrooms, water chestnuts, Chinese cabbage, or snow peas. Shredded almonds, and cooked sliced chicken or pork can also be added, as can cooked chow mein noodles.

More for Your Interest
According to Saunders in *Edible and Useful Wild Plants of the United States and Canada*, some Indian peoples enjoyed chewing and eating the sweet sap that exudes from the stems of common reed grass when these are injured or attacked by insects. The sap was compressed into balls and eaten as candy. Sometimes the sap was dissolved by boiling the stalks in water, and made into a sweet, nourishing drink. The Indians of the Mojave Desert in California actually ground the dried stalks into flour, which they placed near a fire until it swelled and turned brown. They then ate the resulting taffy-like mixture.

The stems were often used in Indian basketry. In southern British Columbia, they were picked while still green, cured in the smoke from an open fire, then used as a decorative overlay in coiled baskets. The mature stems are so strong they were sometimes used by Indians as arrowshafts for hunting birds.

Common Cat-Tail

(Cat-Tail Family)

Other Names
Bulrush, flag, reed-mace, cat-o'-nine-tail, "Cossack asparagus".

How to Recognize
The cat-tail is one of our best-known marsh plants, growing from 1 to 3 m tall and spreading widely by means of perennial underground rootstocks, or rhizomes. The leaves, which grow in sheathing clusters from the base of the plant, are long, erect, and flat, up to 3 cm or more broad, smooth-edged, rounded or pointed at the tips, and filled with spongy pith. The flower stems are long, straight, and pithy. They rise from the centre of the leaf clusters and each bears, at the tip, a cylinder of tightly clustered, greenish female flowers topped by a yellowish spike of male, or pollen-bearing, flowers. The female flowers ripen into a velvety, chocolate-brown seedhead, the well-known "cat-tail", composed of thousands of tiny individual seeds, each attached to a cluster of minute hairs that functions as a parachute once the seed is released, usually in late fall or winter. If the seedheads are taken indoors as decorations, they should be sprayed with varnish or hairspray, as they may "explode" when they dry out, and release their cottony seeds throughout the house.

Another species, the narrow-leaved cat-tail (*T. angustifolia* L.), occurs in eastern Canada and may be used in a similar manner to the common cat-tail.

Where to Find
Cat-tail is common in marshy ground and shallow, standing water along the margins of lakes and streams from British Columbia to Newfoundland, extending northwards into the Yukon. The narrow-leaved cat-tail occurs in southern Ontario and Quebec.

How to Use
Cat-tail is one of the most important and versatile of all edible wild plants. It is very common and in many places can be harvested in large quantities without detriment to the plant populations. The rootstocks, young shoots, and young male flower spikes are all edible. The shoots, gathered in April and May when they are just appearing above the water, are especially delectable and, in our opinion, it would be hard to find a more succulent, easily prepared salad vegetable.

Typha latifolia L.
(Typhaceae)

When pulled, the shoots will break off just where they attach to the rootstocks. The mud-coated outer leaves are simply peeled off and discarded, as are the fibrous upper portions of the leaves. The tender "hearts" are then washed and cut in pieces, to be served raw with a favourite dressing. They can also be simmered gently in water, drained, and served hot with butter.

The long, thick, brown-skinned rootstocks, or rhizomes, are rich in starch and sweetish in taste, but are more difficult to harvest because they must be dug up from the mud. They can be dug from autumn through to early spring and eaten raw or as a cooked vegetable. The central core is somewhat tough and pithy but can be easily removed after the rootstocks are cooked. A flour made by drying the rootstocks until hard, then grinding them to a powder, can be used to make cereals, breads, muffins, and as a thickening for soups and stews. The pure white, succulent growing tips and side-shoots of the rootstocks are a special treat if enough can be gathered. Peeled and eaten raw they are like celery. They can also be sautéed or steamed and make an excellent side dish with meat.

The young flower spikes (male and female) are a palatable vegetable while still green and immature, and can be eaten raw, boiled, or steamed, or used as an ingredient in soups, stews, and casseroles. However, the central core, or axis, of the male spike is tough and must be discarded. The spikes are ready to pick any time from early June to August, depending on the locality, just when the light-green leaf sheath encasing the flower head begins to split. The taste of the young spikes has been compared variously with that of olives, globe artichokes, and sweet corn.

Cat-tail pollen itself, when ripe, can be gathered by shaking the pollen spikes into a bag. The pollen can be substituted for up to half the required flour in pancakes, biscuits, or muffins, giving these foods a unique flavour and bright-yellow colour.

Warning
Do not confuse cat-tail with wild iris, or flag (*Iris* species), which grows in the same habitat as cat-tail and has poisonous leaves and rootstocks, according to Kingsbury in *Poisonous Plants of the United States and Canada*. When in flower, wild iris may be easily recognized from the similarity of its flowers, whether yellow, white, or blue, to those of garden iris. Recognition is more difficult with vegetative plants; the leaves are linear, erect, parallel-veined, and two-ranked, and are usually not as tall as those of cat-tail. The rhizomes are fleshy and horizontal, and usually thicker than those of cat-tail.

Be careful not to collect cat-tails, especially roots and new shoots, from water polluted by sewage or industrial wastes. Always wash your harvest thoroughly in clean, fresh water.

wild iris (*Iris pseudacorus*)
Poisonous: do not eat
see Warning

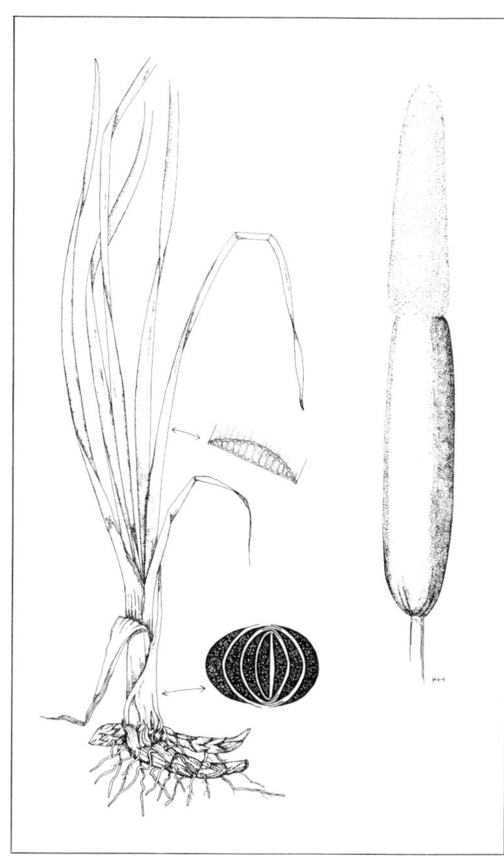
cat-tail (*Typha latifolia*)

Suggested Recipes

Cat-Tail–Cashew Salad

500 mL	young cat-tail shoots	2 cups
125 mL	roasted, salted cashew nuts	1/2 cup
30 mL	salad oil	2 tbsp
15 mL	wine vinegar *or* lemon juice	1 tbsp
	salt, black pepper	

Peel and wash cat-tail shoots and cut into 2-cm lengths. Add cashews. Pour over oil and vinegar *or* lemon juice. Season to taste and toss lightly but thoroughly. Serve at once. Serves 2–3.
Note: hazelnuts or sunflower seeds can be substituted for cashews.

Cat-Tail Flower-Spike Casserole

	about 2 dozen young cat-tail flower spikes, male and female	
500 mL	whole-wheat bread-crumbs	2 cups
30 mL	softened butter *or* margarine	2 tbsp
5 mL	tarragon	1 tsp
5 mL	fresh sage	1 tsp
	or	
2 mL	dried sage	1/2 tsp
	1 small onion, finely chopped	
2 mL	salt	1/2 tsp
	1 egg, well beaten	
50 mL	milk	1/4 cup

Place cat-tail spikes in a saucepan, barely cover with water, bring to a boil, and simmer until spikes are tender (5 to 10 minutes). Scrape off the fleshy outer portions and discard the hard inner cores. In another bowl mix breadcrumbs with softened butter *or* margarine, add herbs, onion, and seasoning, and mix thoroughly. Add crumb mixture to cat-tail scrapings and place in a greased casserole. Beat together the egg and milk and pour over the mixture. Bake at 160°C (325°F) for 30 minutes, or until the egg sets and top is browned. Serves 2.

Cat-Tail "Corn"

Gather about 2 dozen young cat-tail, male and female, flower spikes, cover with salted water in a saucepan, bring to a boil, and simmer for about 10 minutes. Serve, like tiny ears of corn on the cob, with melted butter, salt, and pepper. The fleshy part can be chewed off the "cob" or cut off with a knife. Children will love this simple dish. Serves 3–4.

Cat-Tail Rhizome Flour

Collect and wash about 2 dozen 25-cm (10 in.) long sections of cat-tail rhizome. Peel off the brown skin with a sharp knife, lay peeled sections on a cookie sheet, and dry them in a slow oven (about 100°C or 200°F) for about 1 hour, or until dry and brittle. (Or you can spread them out in the sun for several days until the dry, brittle stage is reached.) Then break them into small pieces and place in a food grinder or blender and grind into flour. Sift to remove any lumps or fibres. You should obtain about 500 mL (2 cups) cat-tail flour, which can wholly or partially replace ordinary wheat flour in any baking recipe. Cat-tail flour is reported to contain 30 to 46 per cent starch, and about the same amount of protein (6 to 8 per cent) as rice and corn flours, but with less fat, according to Harrington in *Edible Native Plants of the Rocky Mountains*. Harrington cites one early researcher, who estimated that 2.5 hectares (1 acre) of cat-tails would yield about 2937 kg (6475 lb) of flour.

Cat-Tail Flour Biscuits

250 mL	cat-tail rhizome flour	1 cup
250 mL	unbleached wheat flour	1 cup
25 mL	baking powder	1 1/2 tbsp
3 mL	salt	3/4 tsp
75 mL	shortening	1/3 cup
175 mL	milk	3/4 cup

Blend together cat-tail and wheat flour and sift with baking powder and salt. Cut in shortening until mixture is fine and crumbly. Add enough milk to make a soft dough. Turn onto a floured board and knead lightly, then roll to 1 cm thickness. Cut with floured biscuit cutter, place on greased cookie sheet, and bake at 230°C (450°F) for 10 to 12 minutes, until biscuits are firm and lightly browned. Makes about 1 dozen.

More for Your Interest
The Cossacks of the Don River marshes of the U.S.S.R. eat the young shoots of the cat-tail and have used it for centuries as a raw vegetable, hence the name "Cossack asparagus".

The tough, spongy leaves of the cat-tail have been used the world over for weaving mats, baskets, hats, slippers, and even chair-seats. Western North American Indians used cat-tail mats for constructing their temporary summer houses, as insulation for winter houses, and for sleeping mattresses. The leaves were not woven but sewn together with thread from the base of the cat-tail leaves themselves or from some other plant fibre. A long, sharp, wooden needle was used to sew the leaves, which were laid out side by side, alternating upper and lower ends. You can make attractive Indian-style placemats from cat-tail leaves by sewing them with tough thread and a large darning needle. Pick the leaves in late summer, dry them thoroughly in the sun, then, just before using, soak them in warm water until pliable. The Indians also used the fluffy seedheads as infant bedding and diapers. This cottony material can also be used to stuff pillows and mattresses and as tinder for starting fires. The mature fruiting heads can be soaked in kerosene to make effective torches.

Cat-tails provide an important refuge for many forms of wildlife. Geese and muskrats feed on the starchy rootstocks, and Blackbirds and other songbirds regularly nest in cat-tail swamps.

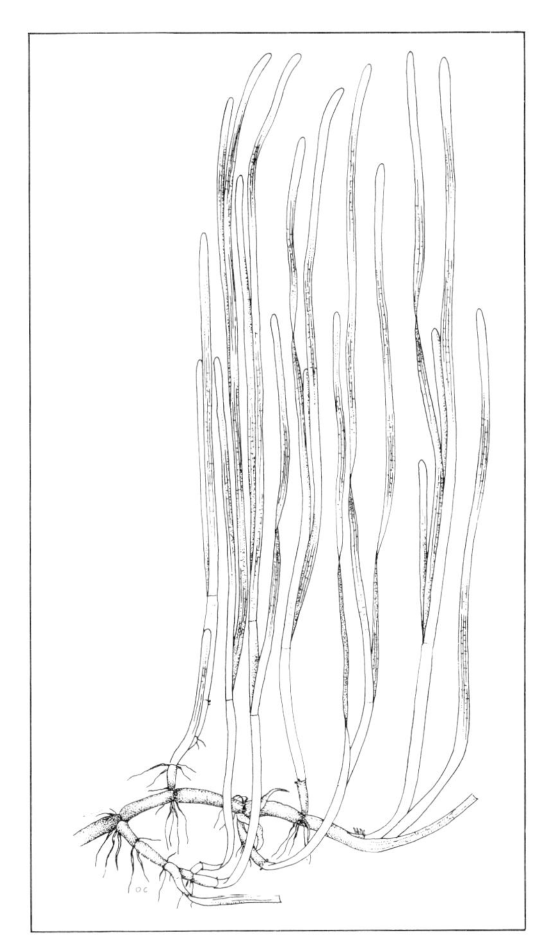

Eelgrass

(Eelgrass Family)

Other Names
Grass-wrack, sea-grass.

How to Recognize
This is a grass-like, marine, flowering plant that grows from brownish-to-whitish creeping rootstocks, or rhizomes. The pale-green, ribbon-like leaves can reach lengths of 3 m or more, but usually are around 1 m long and 3 to 12 mm broad. The inconspicuous flowers and fruits are enclosed within sheaths at the bases of some of the leaves.

Where to Find
Eelgrass is found along both Atlantic and Pacific coasts, where it grows in quiet, protected bays, with rhizomes embedded in mud or sand. Often the plants form dense, extensive subtidal beds, from about 1 m above the lowest tide level to 6 m below.

How to Use
The whitish leaf bases, stems, and crisp rootstocks are all edible and have a refreshing sweetish salty flavour. The upper leaves are usually too tough to eat and must be broken off and discarded. The rootstocks may be muddy or sandy and should be washed thoroughly. The best time to gather eelgrass is

***Zostera marina* L.**
(Zosteraceae)

during a low tide in late spring or early summer when the shoots are young and tender. The plants can be pulled up by hand or dug out with a small trowel or, if the water is too deep, they can be gathered from a boat or canoe with a long "twisting" pole, which can be used to entangle the leaves and uproot the plants. This latter method was commonly used by the coastal Indians of British Columbia, who used eelgrass as a spring vegetable.

The leaf bases and rootstocks are best eaten raw, as a nibble or salad vegetable. The Northwest Coast Indians enjoyed them tied in small bundles and dipped in seal oil or eulachon grease. You can also try steaming eelgrass and serving it with butter and salt as you would fresh green beans. Eelgrass is an excellent vegetable to cook with shellfish and other seafood and is also a good survival food for those lost or stranded along the seacoast.

Warning

Do not collect eelgrass around populated areas, where the water may be polluted by sewage or industrial wastes.

Suggested Recipes

Eelgrass, Kwakiutl Indian Style

Spread washed eelgrass rootstocks, with stems and leaf bases attached, on mats, 1 dozen in front of each person. Each person should remove the small roots from the rootstocks, peel off the outer leaves, then tie the eelgrass plants together in bundles of four, using the leaf bases for tying. The plants in each bundle should be the same length. Each bundle is then dipped in eulachon grease *or* other oil and eaten with the fingers.
Note: the Kwakiutl Indians of northeastern Vancouver Island ate eelgrass prepared and served as above at large tribal feasts. Leftovers could be taken home to the guests' wives. Custom declared that water should not be drunk after such a feast. An eelgrass feast was important, because it was believed that eelgrass was the food of the mythical ancestors of the Kwakiutl people.

Sautéed Eelgrass with Chicken

15 mL	cooking oil	1 tbsp
125 mL	chicken breast, cut into small pieces	1/2 cup
500 mL	eelgrass rootstocks, stems, and leaf bases cut into short lengths	2 cups
250 mL	cold water	1 cup
15 mL	corn starch	1 tbsp
15 mL	soy sauce	1 tbsp
2 mL	fresh ground ginger	1/2 tsp
50 mL	sesame seeds	1/4 cup

Heat cooking oil in a skillet or wok until it bubbles vigorously when a drop of water is added. Add chicken and cook, stirring rapidly, until flesh is white and firm. Add eelgrass pieces and stir rapidly for about 1 minute. Mix together water, corn starch, and soy sauce and add to skillet along with ginger. Continue cooking and stirring until liquid thickens and becomes clear and eelgrass is tender but still crisp. Serve sprinkled with sesame seeds. Good with rice and other Oriental dishes. Serves 2–3.

More for Your Interest

Eelgrass was a favourite food of the coastal Indians of British Columbia. In some Indian languages the name for eelgrass is derived from the word for "sweet". Eelgrass leaves were also used by the Northwest Coast peoples to gather herring eggs in spring. The spawn would be scraped off and eaten raw or cooked; the leaves themselves were not eaten. The people of the Outer Hebrides Islands of the British Isles chewed eelgrass rootstocks when the plants were washed up on the beach after a storm.

During the 1930s a serious disease eliminated nearly all the eelgrass populations along the Atlantic coast. (Pacific eelgrass populations were not affected.) Recently, however, the plant has made a good comeback and may again become as plentiful as it was previously. Many types of waterfowl, including Mallards, Pintails, Scaups, Scoters, Widgeon, and Canada Geese, feed on the seeds, leaves, and rootstocks of eelgrass. For Brant Geese the plant is a staple food, and many American Brant suffered from starvation after the decimation of the Atlantic eelgrass beds.

Cow Parsnip

(Celery Family)

Other Names
"Indian rhubarb", "Indian celery".

How to Recognize
Cow parsnip is one of the many edible plants of the celery family; others include parsnip, carrot, celery, parsley, angelica, dill, and many of our favourite spices, such as caraway and cumin. There are also some highly poisonous plants in this family, such as poison hemlock and water hemlock (see Warning below).

This species of cow parsnip is a robust, hollow-stemmed, herbaceous perennial, up to 2 m or more tall, growing from a stout taproot. The lower leaves are large and compound, each divided into 3 coarsely toothed, sharply lobed segments. The upper leaves are simple but sharply lobed, with conspicuously winged stems. The flowers are borne in large, dense, umbrella-like clusters, or umbels, each with 15 to 30 rays. The individual flowers are small and white. There are often several flower umbels per plant, but the terminal one is largest, and may exceed 25 cm across. The ripe fruits are brownish, papery, and flattened, with thin lateral wings. When mature, the leaves and stems have a strong, pungent odour.

Heracleum lanatum **Michx.**
(Apiaceae or Umbelliferae)

Where to Find

Cow parsnip is found from coast to coast in Canada except in the Arctic tundra zone. It grows in moist, open areas, such as along roadsides and in meadows, from sea level to above the tree line in the mountains, often forming extensive patches.

How to Use

The tender young budstems and leafstalks are mild and sweet and make excellent vegetables, either raw or cooked. In taste and appearance they closely resemble celery and can be used as a celery substitute in most dishes. The stems should always be peeled before being eaten, as the outer skin has a pungent smell and contains a skin-irritating chemical (see Warning). The ideal time to eat the stalks is usually in April or May, before the flower buds have expanded, when they are still enclosed in the broad, sheathing leaf bases. After flowering, the stems become tough and sharp tasting and must be boiled in two or three changes of water to make them palatable.

Cow parsnip stalks can be served in salads or cut into sticks, as you would celery or carrots. Being hollow, they can be stuffed with fillings and spreads such as cream cheese, shrimp, or mushroom. They can also be cooked in stews or stuffed with sausage meat, rice, or cheese filling, and baked in the oven.

As cow parsnip is abundant, widespread, and very palatable when young, it is a highly significant wild food and survival food for the spring and summer months.

poison hemlock (*Conium maculatum*)
Poisonous: do not eat
see Warning

water hemlock (*Cicuta* species)
Poisonous: do not eat
see Warning

Warning

Be careful not to confuse cow parsnip with the highly toxic water hemlock (*Cicuta* species) or the weedy poison hemlock (*Conium maculatum* L.), both of which can cause sickness and death in humans and animals. These plants are more slender than cow parsnip, with smaller, more numerous flower heads and finely divided leaves. Water hemlock tends to grow in standing water in swamps and along creeks and lake margins. Poison hemlock grows in disturbed ground, usually near human settlements.

Always peel cow parsnip stalks, because the outer skin contains a chemical, furanocoumarin, which reacts with the skin and lips, causing blistering and browning in the presence of long-wave ultraviolet light. Harvesters of this plant are advised to wear gloves and to keep skin contact with the plant at a minimum until the shoots and leafstalks are peeled. Once peeled, too, the stems lose much of their pungent odour. Children should avoid playing where these plants grow.

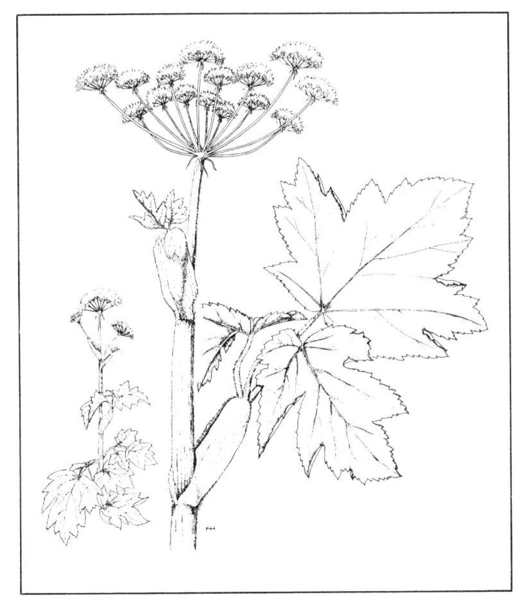

cow parsnip (*Heracleum lanatum*)

Suggested Recipes

Cow Parsnip–Mushroom Salad

250 mL	young cow parsnip stalks, peeled and diced	1 cup
250 mL	fresh mushrooms, sliced	1 cup
250 mL	cold cooked potatoes, diced	1 cup
30 mL	wine vinegar	2 tbsp
30 mL	olive oil	2 tbsp
	salt and pepper	
50 mL	chopped walnuts (optional)	1/4 cup

Mix together cow parsnip, mushrooms, and potatoes. Dress with vinegar and oil and season to taste. Garnish with walnuts if desired. Serves 3–4. Other dressings, such as caper or blue cheese, can be used instead of oil and vinegar.

Cow Parsnip with Anchovy Dressing

	1 dozen young cow parsnip stalks about 10 to 15 cm long, peeled	
	6 anchovy fillets	
30 mL	olive oil	2 tbsp

Split cow parsnip stalks lengthwise. Mash together anchovies and olive oil and use as a filling for the stalks, as you would for celery sticks. Chill and serve as an appetizer. Serves 4–6.

Cow Parsnip Salad

500 mL	fresh bean sprouts	2 cups
250 mL	young cow parsnip stalks, peeled and diced	1 cup
500 mL	cooked pork, diced	2 cups
	2 medium-sized green onions, finely chopped	
	1 large carrot, grated	
50 mL	roasted almonds *or* macadamia nuts	1/4 cup
30 mL	soy sauce	2 tbsp
	salt and pepper	
125 mL	sour cream	1/2 cup
	dash of paprika	

In a large salad bowl, mix together sprouts, cow parsnip, pork, onions, carrot, nuts, and soy sauce. Season to taste and dress with sour cream sprinkled with paprika. Serve with steamed rice. Serves 4–6.

Cow Parsnip with Oysters

250 mL	young cow parsnip stalks, peeled and finely diced	1 cup
30 mL	butter	2 tbsp
	1 dozen medium-sized oysters	
250 mL	semi-sweet white wine	1 cup
	salt and pepper	
	pinch of garlic salt	

In a skillet sauté cow parsnip in butter until tender but not mushy. Add oysters and liquor from oysters and simmer until the edges of the oysters curl. Add white wine and seasoning and heat until steaming. Serve immediately. Serves 3–4. As a variation, sherry can be substituted for the wine, and the oysters and cow parsnip can be served over toast.

More for Your Interest

Cow parsnip was used as a vegetable by many Indian peoples of Canada and the United States. The coastal Indians of British Columbia often ate the shoots with seal oil, dogfish oil, or eulachon grease and, within the last century or so, enjoyed them sprinkled with sugar or dipped in honey.

Various species of *Heracleum* are eaten in many parts of the world, from the Hebrides in Scotland to Chile in South America. The Prussians cultivate one species (*H. sibiricum* L.) as an animal fodder, especially for ewes. In Lithuania and parts of Russia, cow parsnip is distilled, alone or mixed with bilberries, to make a popular alcoholic beverage.

Cow parsnip has considerable horticultural potential, both as a vegetable and as an ornamental, as it has very attractive foliage and flowers. Swallowtail butterfly larvae feed on the leaves.

Indian Celery

(Celery Family)

Other Names
Indian consumption plant, desert parsley.

How to Recognize
"Indian celery" is a taprooted perennial with single or clustered, erect stems up to 1 m high, but usually under half a metre. The foliage is smooth, bluish-green and waxy in appearance. The leaves are 1 to 3 times compound, with 3 to 30 well-defined oval to lance-shaped leaflets, each 2 to 9 cm long. These are either smooth-edged or toothed at the tips. The leafstalks are winged and clasp the stems. The small, pale-yellow flowers are borne in tight balls at the ends of several to many irregularly spaced rays that form an umbrella-like cluster, or umbel. The umbel rays elongate as the flowers and fruits mature; the longer ones can exceed 20 cm. The fruits are dry, oblong or elliptic, and flattened, with broad, light-coloured, lateral wings. The entire plants, and especially the fruits, when crushed, have a strong, celery-like aroma, due to the presence of an aromatic oil.

Almost all of the many other lomatiums in western Canada can be used as root vegetables or greens, but this species is one of the most widespread and plentiful and also one of the tastiest. Most of the other lomatiums have more finely dissected, feathery leaves.

***Lomatium nudicaule* (Pursh) Coult. & Rose**
(Apiaceae or Umbelliferae)

Where to Find

This species grows on grassy bluffs and in dry, open meadows and woods from sea level to moderate elevations in the mountains, from southern coastal British Columbia to south-western Alberta. Lomatiums do not occur in eastern Canada.

How to Use

The shoots, leafstalks, and leaves of this plant have a strong, though pleasant, celery flavour and aroma and, when young, are delicious eaten raw, cooked as a potherb, or used to flavour soups and stews in the same way one would use celery leaves. As with most leafy vegetables, the young leaves are the best for eating. Since the leaves become stronger-tasting as they get older, it is best to gather them before the plants go to seed. After light blanching, the leaves can be canned or frozen.

The fruits can also be used as flavouring, in the same way one would use celery seed, fennel, or caraway seed. The plants can be used fresh or dried as a tea. Simply place a handful of the leaves (fewer are required if dried) in about a litre of water, bring to a boil, cover, and allow to steep a few minutes before serving.

Suggested Recipes

"Indian Celery"–Smoked Salmon Salad

500 g	smoked salmon	1 lb
500 mL	young "Indian celery" leaves and leafstalks, cut into bite-sized pieces	2 cups
125 mL	mayonnaise	1/2 cup
	salt and pepper	
	2 hard-boiled eggs, sliced	
	pitted ripe olives (optional)	

Break salmon into bite-sized chunks and place in salad bowl along with "Indian celery", mayonnaise, and seasoning. Mix well and chill. Serve garnished with hard-boiled egg slices and sliced olives, if desired. Serves 4.

Butter-Steamed "Indian Celery" with Cheddar Cheese

50 mL	butter *or* margarine	1/4 cup
500 mL	young "Indian celery" leaves and leafstalks, shredded	2 cups
50 mL	water *or* meat broth	1/4 cup
125 mL	Cheddar cheese, grated	1/2 cup
	dash of paprika	
	salt and pepper	

In a heavy skillet, melt butter *or* margarine. Add "Indian celery" and water *or* broth and stir, then cover and allow to simmer for 5 minutes, or until tender. Season to taste and serve hot, sprinkled with grated Cheddar cheese. Serves 3–4.

"Indian Celery" with Ham

	8 slices bacon	
50 mL	beef *or* chicken broth	1/4 cup
250 mL	ground, cooked ham	1 cup
	1 large onion, chopped	
500 mL	young "Indian celery" leaves and leafstalks, shredded	2 cups
	6 slices buttered toast	
250 mL	Swiss cheese, grated	1 cup
	salt and pepper	

In a large skillet, fry bacon until crisp, then drain, and put aside. Return 30 mL (2 tbsp) bacon drippings to pan, add broth, ham, onion, and "Indian celery" and cook gently until celery has softened (about 5 minutes). Remove from heat and add crumbled bacon. Arrange toast on broiler pan and spoon "Indian celery" mixture overtop. Sprinkle grated cheese over, season to taste, and place under broiler until cheese melts. Serve hot. Serves 3–4.

More for Your Interest

The Interior Salish Indians of British Columbia enjoyed this plant as a green vegetable, and still eat it today. "Indian celery" tea was drunk as a tonic for colds and sore throats and the seeds were widely used as a fumigant and air-purifier, a flavouring for tobacco, and a medicine. To purify the air, a few seeds were placed in a fire or on top of a hot stove and allowed to smoke until their odour had permeated the entire house. As a medicine for colds, sore throats, and tuberculosis, the seeds were chewed and the juice swallowed. The pulverized seeds were plastered over skin sores as a poultice. The seeds were also used as a valuable trading item among the Indians of southern British Columbia.

"Indian celery" propagates well from seed and can be an interesting, attractive, and useful addition to any herb garden.

Balsamroot

(Aster or Composite Family)

Other Names
Spring sunflower, "wild sunflower".

How to Recognize
This perennial plant grows up to half a metre high from a thick, tough, deep-seated taproot. The leaves are large, triangular, and long-stalked, rising in a dense cluster from the root-crown. A thick covering of fine, white hairs, especially prominent on the undersides, gives the leaves a silvery caste. The showy flower heads with their bright-yellow rays, 15 to 25 per flower, are like small sunflower heads. They are borne on individual stems, usually many per plant. The small, black, wedge-shaped fruits shake loose easily from the old, dry heads. The flowers bloom in early spring and are very prolific, particularly on grazed hillsides. In many places within their range they are so abundant that they colour entire hillsides a golden yellow. A second species, also edible, *B. deltoidea* Nutt., is similar in appearance to *B. sagittata*, but has brighter-green leaves.

Where to Find
Balsamroot grows on dry, sunny hillsides and in open woods throughout the dry belt of southern British Columbia and in the Alberta foothills. *B. deltoidea* occurs in rocky outcrops on the extreme southern coast of British Columbia, and on Vancouver Island.

How to Use
Balsamroot is one of the most versatile of edible wild plants. The thick taproots are edible in spring and were an important food of the Interior Indians of British Columbia.

***Balsamorhiza sagittata* (Pursh) Nutt.**
(Asteraceae or Compositae)

They also make an interesting coffee substitute (see the second publication in the Edible Wild Plant series, *Wild Coffee and Tea Substitutes of Canada*). The seeds can be ground and eaten. In our opinion, however, the best of all the edible parts of this plant are the tender young flower budstems. These should be picked off in spring, when about 10 cm long, while the buds are still tight and green. They snap off easily at the base, like asparagus shoots. The outer skin is easily peeled off and discarded and the succulent inner stems make a delicious, nut-flavoured nibble or cooked green vegetable, with a flavour similar to that of Jerusalem artichokes. Be sure to pick only a few budstems from each plant, in order to leave enough to bloom.

Even earlier in the spring, before the buds or leaves appear above the ground, you can locate the plants from the dead flowerstalks of the previous year and dig down to obtain yet another edible part of this plant—the delicate white mass of unsprouted leaf—and flower buds. These can be cut off from the root-crown and eaten raw or lightly cooked. In maturity, the leaves and flower heads of balsamroot become very tough and strong-tasting, and are inedible.

Suggested Recipes

Balsamroot Budstem Snacks

	2 dozen young balsamroot flower budstems, 10 cm long	
30 mL	cream cheese	2 tbsp
30 mL	mayonnaise	2 tbsp
	dash of paprika	
1 mL	celery salt	1/4 tsp
	salt to taste	

Remove and discard flower buds, peel stems and arrange on a tray. In a small dish, mix together remaining ingredients until smooth and creamy. Dip balsamroot stems and eat as a snack or hors d'oeuvre. Serves 4–6.

Steamed Balsamroot Budstems

	3 dozen tender, young balsamroot flower bud-stems	
15 mL	butter *or* margarine	1 tbsp
	dash of nutmeg	
	salt and black pepper	
	1 lemon, halved	

Remove and discard flower buds, then peel and cut stems into 2-cm lengths. Place in a food steamer or in a saucepan with about 50 mL (1/4 cup) water. Steam or boil about 5 minutes, or until just soft. Drain and serve with butter *or* margarine and a sprinkling of nutmeg, salt, and pepper. Squeeze juice from one half the lemon overtop and arrange half-slices from the other half around the edges of the dish as a garnish. Serves 2.

More for Your Interest

The large, woolly leaves of balsamroot, some as much as 30 cm long and 15 cm wide, were used by the Okanagan Indians of British Columbia as insoles in their moccasins, to keep the feet warm. Young Okanagan boys would test their skill in careful footwork by wrapping the leaves on their feet, securing them with grass stems, and walking on them to see how far they could go without tearing them. Some sure-footed youths were even able to run, still keeping the leaves intact.

The mature leaves are unpalatable to animals, but the young leaves are grazed by deer, elk, mountain sheep, and domesticated stock.

Rock-Cresses

(Mustard Family)

Arabis alpina

Other Names
Cress, upland cress.

How to Recognize
There are at least fifteen species of *Arabis* native to Canada and several Eurasian species that have been introduced as rock-garden flowers and have escaped cultivation in some localities. All of these are edible but some are milder tasting and less hairy than others and thus more suitable for eating. In general, those species with densely woolly or hairy leaves are good only after being cooked, and even then, are inferior to the smooth-leaved species.

***Arabis* species**
(Brassicaceae or Cruciferae)

As a group, the rock-cresses are biennial (occasionally annual) or perennial herbs with several to many basal leaves that are usually stalked and smooth-edged to variously lobed. One to several erect stems arise from the root-crown. These bear smaller, usually unstalked leaves and a terminal cluster of white, yellowish, or pinkish-purple flowers. In most cases the flowers are quite small and not very conspicuous. The fruits are narrow and elongated, containing many flattened seeds. In some species the fruits are erect and clasping the stem; in others they spread outwards or downwards. Hairs that do occur on the leaves and stems can be simple, forked, or even star-shaped.

Two species, *A. alpina* L. and *A. lyrata* L., have been specified in reference literature as being particularly good to eat. *A. alpina* is a mat-forming perennial up to 40 cm tall, with slightly hairy, shallowly toothed leaves, whitish petals and ascending fruits. *A. lyrata*, a tufted perennial, has smooth or slightly hairy, deeply lobed or toothed leaves, whitish petals, and spreading fruits.

Where to Find

The rock-cresses are aptly named, for they commonly grow on rocky or gravelly slopes. Some species prefer moist ground, some dry, sandy ground. Various species are found throughout Canada; a number grow in alpine and subalpine areas and in the Far North. *A. alpina* occurs in northeastern Canada, from Manitoba to Newfoundland. Varieties of *A. lyrata* occur in many upland localities, from British Columbia to Quebec, north to the Yukon and Northwest Territories.

How to Use

The rock-cresses all have the typical, rather sharp flavour of many plants in the mustard family. Some have leaves that are too small or too hairy to be eaten, but many, including the two specified here, are quite suitable for use in sandwiches and salads and cooked, as potherbs. The tender, young basal leaves and even the flowers can be used. Some people are particularly fond of the hot radish-like flavour of these plants and will enjoy them served alone. Others will prefer to use them more as a flavouring, mixed with other milder-tasting foods. In general, they can be prepared and served in the same manner as watercress or garden cress. They are especially good in combination with cheese or eggs.

Suggested Recipes

Rock-Cress Salad with Cottage Cheese Dressing

500 mL	young rock-cress leaves	2 cups
50 mL	green onion, chopped	1/4 cup
50 mL	green pepper, chopped	1/4 cup
125 mL	cottage cheese	1/2 cup
50 mL	sour cream	1/4 cup
30 mL	lemon juice	2 tbsp
	dash of garlic salt	
	dash of paprika	
	salt and pepper	
	ripe olives, sliced (optional)	

Wash rock-cress leaves and place in salad bowl with green onion and green pepper. Prepare dressing by combining remaining ingredients, except olives, in an electric blender and whirring until smooth. Pour over greens, toss lightly, and garnish with ripe olives. Serve chilled. Serves 4.

Rock-Cress with Bacon

500 mL	young rock-cress leaves	2 cups
	4 slices bacon	
30 mL	sour cream	2 tbsp
	dash of onion powder	
	salt and pepper	

Wash rock-cress leaves, drain, and set aside. Cut bacon into small pieces and fry until crisp. Drain off all but 15 mL (1 tbsp) drippings and add rock-cress leaves to the pan. Add sour cream and seasoning and cook over low heat, stirring lightly until rock-cress leaves are tender (2 to 3 minutes). Do not let the mixture get too hot, otherwise the sour cream might curdle. Serve with eggs for breakfast. Serves 2–4.

More for Your Interest

The Eskimos of Alaska ate the young leaves of *A. lyrata* raw or boiled and also liked to ferment them for winter use.

The genus name *Arabis* was derived from the country, Arabia, where rock-cresses are found. Several *Arabis* species are cultivated in rock gardens at present, and a number of wild ones should be considered for use in horticulture.

Sea-Rocket

(Mustard Family)

How to Recognize

Sea-rocket is a simple or branching annual up to half a metre or more tall, with spreading lower stems and erect upper stems. The leaves are fleshy, simple, and alternate with oblong oval-shaped blades, usually under 5 cm long, deeply wavy or irregularly lobed at the edges, narrowing at the base. The 4-petalled flowers are inconspicuous, white to purplish-tinged, and grow in loose clusters at the ends of the stems. The fruits are fleshy and 2-jointed, the oval, pointed, upper joint detaching at maturity.

Where to Find

Sea-rocket is found in sandy or gravelly soil along the Atlantic coast, in southern Labrador, Newfoundland, and the Maritimes, and in some localities around the shores of the Great Lakes. It also occurs in abundance along the Pacific coast, where it is thought to have been introduced.

***Cakile edentula* (Bigel.) Hook.**
(Brassicaceae or Cruciferae)

How to Use

The young, succulent leaves of this plant have a sharp mustard flavour, similar to that of horseradish. They are excellent in salads or as a sandwich green. Some people prefer to make salads consisting wholly of sea-rocket leaves with perhaps a small amount of onion. Others like to mix the leaves half and half with some milder green, such as lettuce or spinach. The fleshy green fruits are also edible and both leaves and fruits can be pickled, like capers or nasturtium seeds, in vinegar.

Sea-rocket leaves can be harvested from May through the summer months. If you live near a patch of sea-rocket, you can maintain a continuous supply of fresh, young leaves simply by harvesting the growing tips, which will regenerate within one or two weeks. If you stagger your picking over many plants you will not deplete the populations. Be sure to allow some plants to flower and produce seed before the summer finishes, however, as this plant is an annual and must produce seed to grow the following year.

Suggested Recipes

Scrambled Eggs with Sea-Rocket

250 mL	young sea-rocket leaves	1 cup
	1 medium onion, thinly sliced	
15 mL	butter *or* margarine	1 tbsp
	salt and pepper	
	6 large eggs	
125 mL	light cream	1/2 cup

Wash, drain, and chop sea-rocket leaves. Fry onion in butter *or* margarine until soft, then add sea-rocket and seasoning, and sauté over slow heat for about 10 minutes. Mix together eggs and cream and add to pan. Scramble for a few minutes, until eggs set. Serve hot on toast with hot asparagus or other suitable vegetable. Serves 4–6.

Sea-Rocket Salad

500 mL	young sea-rocket leaves	2 cups
250 mL	green pepper, thinly sliced	1 cup
	1 medium onion, thinly sliced	
50 mL	wine vinegar	1/4 cup
50 mL	olive oil	1/4 cup
	salt and pepper	
	sliced tomatoes (optional)	
	black olives, sliced (optional)	

Mix together sea-rocket, green pepper, and onion. Add vinegar, olive oil, salt, and pepper. Garnish with sliced tomato and olives. Serve cold. Serves 4–6.

Crab–Sea-Rocket Salad

	2 large crabs, cooked, cracked, and shelled *or* 1 large can of crab meat	
500 mL	sea-rocket leaves, chopped	2 cups
250 mL	green pepper, finely chopped	1 cup
50 mL	green onion, finely chopped	1/4 cup
	2 large tomatoes, sliced	
30 mL	walnuts, chopped	2 tbsp
125 mL	mayonnaise	1/2 cup
50 mL	lemon juice	1/4 cup
30 mL	chili sauce	2 tbsp
	salt and pepper	
	4 to 6 lettuce leaves (optional)	
	2 hard-boiled eggs, sliced	

Lightly mix together crab meat, sea-rocket, green pepper, onion, tomatoes, and walnuts. Add mayonnaise, lemon juice, and chili sauce. Mix together and season with salt and pepper. Serve on individual plates, on lettuce leaves if desired, and garnish with egg slices and crab legs. Serves 4–6.

More for Your Interest
According to the explorer Pehr Kalm (cited in *Sturtevant's Edible Plants of the World*), who visited Canada in the mid-eighteenth century, the early settlers in Canada used to dig, dry, and pound the roots of sea-rocket to a powder, and mix it with flour to use when there was a scarcity of wheat.

The name *Cakile* derives from *qāqula*, an old Arabic word, probably applied to the Mediterranean species *C. maritima* Scop., which has been introduced sporadically on the Atlantic and Pacific coasts of North America. The term *rocket* is an early name used for a variety of plants in the mustard family. Originally applied to the annual salad plant *Eruca sativa* Mill. of the Mediterranean and western Asia, it was corrupted to *ruchetta* by the Italians, *roquette* by the French, and finally to "rocket" by the English.

Scurvy-Grass

(Mustard Family)

Other Name
Spoonwort.

How to Recognize
Scurvy-grass is a smooth, somewhat fleshy biennial (or rarely perennial). In the first year it forms a rosette, or basal cluster, of spoon-shaped, smooth-edged or shallowly toothed leaves with slender stalks. The small, white flowers appear early in the second season, forming small clusters that are at first hidden among the leaves, but later elongate and become more conspicuous. The fruiting stems are arched or creeping, up to 30 cm long, with a few scattered oblong leaves. The fleshy fruits are globular to elliptic, up to 8 mm long, and borne on slender, spreading stalks. The entire plant has a horseradish-like odour and the flavour of cress.

Where to Find
Being quite salt-tolerant, scurvy-grass grows in moist places near sea beaches, and because it requires high concentrations of nitrogen it thrives near sea-bird nesting sites. It is common along much of Canada's sea coast, from Vancouver Island up the Pacific coast, along the entire Arctic and Atlantic coasts to Labrador, Newfoundland, and the Gulf of St. Lawrence.

***Cochlearia officinalis* L.**
(Brassicaceae or Cruciferae)

How to Use

The crisp, succulent foliage of this plant can be eaten either fresh or cooked. The leaves are quite rich in ascorbic acid (vitamin C): a study by Hoffman *et al.* found the seed pods and stems to contain 111 mg ascorbic acid per 100 g fresh weight and 28 μg β-carotene per gram fresh weight in the sample analyzed.

Scurvy-grass leaves, like watercress, make a good salad. They are also delicious in sandwiches with a meat, cheese, or egg filling, or alone with bread and butter. The leaves should be gathered when young, preferably from the first-year plants. As a potherb they can be simmered lightly in a small quantity of water.

Suggested Recipes

Scurvy-Grass Salad

250 g	smoked sausages, cut in small pieces	1/2 lb
15 mL	butter *or* margarine	1 tbsp
500 mL	fresh scurvy-grass leaves	2 cups
	a dozen watercress sprigs, cut into bite-sized pieces	
	salt and pepper	
125 mL	olive oil *or* other salad oil	1/2 cup
50 mL	wine vinegar	1/4 cup
30 mL	capers (optional)	2 tbsp

In a skillet, brown sausages in butter *or* margarine and drain off fat. Place greens in a bowl, add sausages, seasoning, and dressing made by mixing oil, vinegar, and capers if desired, then toss lightly and serve while sausage is still hot. Serves 3–4. Other dressings such as French dressing are also good with this salad, and anchovies can be substituted for the sausage.

Scurvy-Grass Salad with Rice

125 mL	uncooked brown rice	1/2 cup
250 mL	cold chicken broth	1 cup
500 mL	fresh scurvy-grass leaves	2 cups
	3 medium-sized mushrooms, sliced	
	juice of 1 lemon	
50 mL	mayonnaise	1/4 cup
	dash of thyme *or* tarragon	
	salt and pepper	

In a saucepan add rice to chicken broth, cover, heat to boiling, and simmer about 40 minutes, or until tender. Chill. Meanwhile, mix together other ingredients, and serve over chilled rice. This is a good summertime buffet dish, served with smoked fish. Serves 3–4.

More for Your Interest

Scurvy-grass was one of the first wild greens to be used by early mariners, explorers, prospectors, and traders across the northern part of the continent to combat scurvy, a disease directly linked to vitamin C deficiency. A number of other plants rich in vitamin C are also known as "scurvy grass", but most have alternate names that are more commonly used. It was, and still is, an important vitamin supplement for northern peoples, not only in Canada, but in Scotland and northern Europe as well. It is sometimes cultivated in northern gardens.

The name *Cochlearia* is derived from the Greek *cochlear*, "a spoon", referring to the spoon-shaped basal leaves; hence the alternate common name of spoonwort.

Prickly-Pear Cacti

(Cactus Family)

Opuntia fragilis **(Nutt.) Haw. and**
O. polyacantha **Haw.**
(Cactaceae)

Other Names

O. fragilis is known as brittle prickly-pear and *O. polyacantha* as many-spined prickly-pear.

How to Recognize

There are many species of prickly-pear in North America, but only two whose ranges extend into Canada. Both are prostrate perennials, forming extensive clumps or mats. The green, jointed stems are succulent. The stem joints of *O. fragilis* are small (usually under 5 cm long) and rounded in cross-section, whereas those of *O. polyacantha* are larger (up to 15 cm long) and strongly flattened. In cacti the leaves are modified into sharp spines that effectively conserve moisture and act as protection against would-be browsers. These spines, up to 5 cm long in these two species, are straight with barbed tips. They are borne in clusters of 2 to 10 scattered over the surface of the stems. At the base of each spine cluster are many short splinter-like bristles. The flowers, which appear in late spring, are large and showy. They are yellow or sometimes reddish-tinged with age, with many petals and numerous stamens. The reddish-coloured, dry, spiny fruits are oval shaped, up to 2 cm long in *O. fragilis* and up to 2.5 cm long in *O. polyacantha*. Unlike those of the large prickly-pears of the southwestern United States and Mexico, they are little valued as food.

Another cactus species, the purple cactus [*Mamillaria vivipara* (Nutt.) Haw.], is also found in Canada. It is a cushion-like cactus up to 8 cm high and 3 to 20 cm across, with large, showy, purplish flowers and greenish berry-like fruits, which are sweet and edible when ripe.

Opuntia polyacantha

Where to Find
Both prickly-pear species grow on dry, exposed, rocky or sandy ground. *O. fragilis* occurs in Canada in the southwestern prairies and in British Columbia, in the dry interior region and along the rocky coastline of Vancouver Island and the Gulf Islands. *O. polyacantha* is common throughout the southern arid portions of the Prairie Provinces and also in the southern interior and southern coastal region of British Columbia. The purple cactus is very common on open plains and hillsides throughout the southern prairies.

How to Use
Almost anyone who lives in cactus country has suffered the painful experience of stepping on or brushing against these spiny plants and having to have the barbs and small bristles extracted with pliers or tweezers. It must therefore seem inconceivable to many that these plants should even be considered as a source of food. However, once the spines are removed, either by singeing in a fire or by scalding in water until soft, then being peeled off with a knife, the flesh is quite palatable, comparable in taste to fresh raw cucumber or fresh cooked green beans. The only disadvantage is that the flesh is quite mucilaginous, especially when raw, but this stickiness can be largely eliminated by washing the cut stems in water.

Cactus stems can be eaten in a variety of ways. Raw and fresh they are good in salads or cut up with other vegetables in a cold plate. They are especially good marinated with a little vinegar or lemon juice. When roasted over an open fire, the insides will pop out when squeezed, just like the insides of a roasted marshmallow. The stems are also good in soups and stews, they can be sliced, dipped in batter, and deep-fried, they can be candied, and some Indian people in British Columbia have used them as an ingredient in fruit cakes.

Prickly-pear stems are best when young and fresh in the spring, but can be eaten at any time of the year, even in winter if they can be dug out from under the snow. As such, they are a valuable emergency food. They are high in calcium, phosphorus, and vitamin C.

Suggested Recipes

Broiled Cactus with Canadian Cheddar

500 g	fresh prickly-pear stems	1 lb
30 mL	lemon juice	2 tbsp
30 mL	brown sugar	2 tbsp
50 mL	butter *or* margarine, melted	1/4 cup
50 mL	Canadian Cheddar, grated	1/4 cup
	salt and pepper	

Brush peeled, de-spined stems with lemon juice and sprinkle with sugar. Place on a baking sheet or open casserole dish and broil 5 to 8 minutes, or until tender. Brush with butter *or* margarine, add grated cheese and seasoning and broil an additional 2 or 3 minutes until cheese melts. Serve hot. Serves 3–4.

Pickled Cactus Stems

500 g	fresh prickly-pear stems	1 lb
250 mL	wine vinegar	1 cup
250 mL	brown sugar	1 cup
30 mL	salt	2 tbsp
30 mL	pickling spice, tied in cheesecloth bag	2 tbsp
	6 bay leaves	

Cover peeled, de-spined cactus stems with cold, salted water and let stand for 6 hours. Combine vinegar, sugar, salt, spices, and bay leaves in an enamel pan and bring to a boil. Drop in the stems, bring to a boil again, and simmer 15 minutes. Pour into a sterilized jar, making sure that the stems are covered by the vinegar mixture to within 1 cm from the top. Seal and leave 4 weeks before serving. Makes 1 medium-sized jar.

Prickly-Pear Salad

250 mL	prickly-pear stems	1 cup
	2 green onions, finely chopped	
	2 medium-sized fresh tomatoes, cubed	
50 mL	cider vinegar	1/4 cup
30 mL	olive oil	2 tbsp
2 mL	sugar	1/2 tsp
	salt and pepper	
	4 large lettuce leaves	

Cut peeled, de-spined cactus stems into slices, set on a baking sheet and bake at 160°C (325°F) for about 10 minutes, or until tender. Cool to room temperature, mix with other ingredients, chill 1 to 2 hours, and serve on lettuce with toast. Serves 4.

Prickly-Pear with Lemon Dressing

1 kg	fresh prickly-pear stems	2 lb
5 mL	salt	1 tsp
125 mL	lemon juice	1/2 cup
50 mL	vegetable oil	1/4 cup
15 mL	chives *or* green onions, finely chopped	1 tbsp
5 mL	sugar	1 tsp
	1 clove garlic, minced	
	dash of paprika	
	2 eggs, hard-boiled and chopped	
15 mL	parsley, chopped	1 tbsp

Cover the peeled, de-spined stems with water to which the salt has been added, bring to a boil, and simmer for about 3 minutes. Drain, cool to room temperature, and slice the stems into thin pieces. To make the dressing, combine lemon juice, oil, chives *or* onions, sugar, garlic, and paprika. Add cactus and chill for 1 to 2 hours. Just before serving, garnish with eggs and parsley. Serves 4–6.

More for Your Interest

There are over 130 species of prickly-pear cactus, all native to the western hemisphere. Prickly-pears were being cultivated for their sweet, juicy fruits by the natives of Central America when Europeans first arrived, and were introduced by the Spaniards to the Mediterranean and other parts of Europe. Now they are a major crop in Sicily as well as being cultivated in the southwestern United States and Mexico. The fruits are exported to other parts of the world and can sometimes be purchased as a specialty in Canadian markets.

Eating prickly-pear stems with the spines attached can cause death to cattle and other livestock, but when the spines are rubbed or burned off, the cacti become a valuable and nutritious stock feed. Recently spineless varieties have been successfully cultivated as fodder.

In Canada, prickly-pear cacti were eaten throughout their range by native peoples, and were also placed around the poles of raised food caches to keep away mice and other rodents. The spines were sometimes used to make fish-hooks. The Plains Indians and the early pioneers used the stems to clarify muddy water for drinking. The stems were cut open to expose the mucilaginous insides, then dropped into a container of murky water. Clay and silt particles suspended in the water would gradually become entrapped by the sticky mucilage and the water would become clear enough to drink. Cactus mucilage was also used as a fixative for dyes and paints.

Seabeach-Sandwort

(Pink Family)

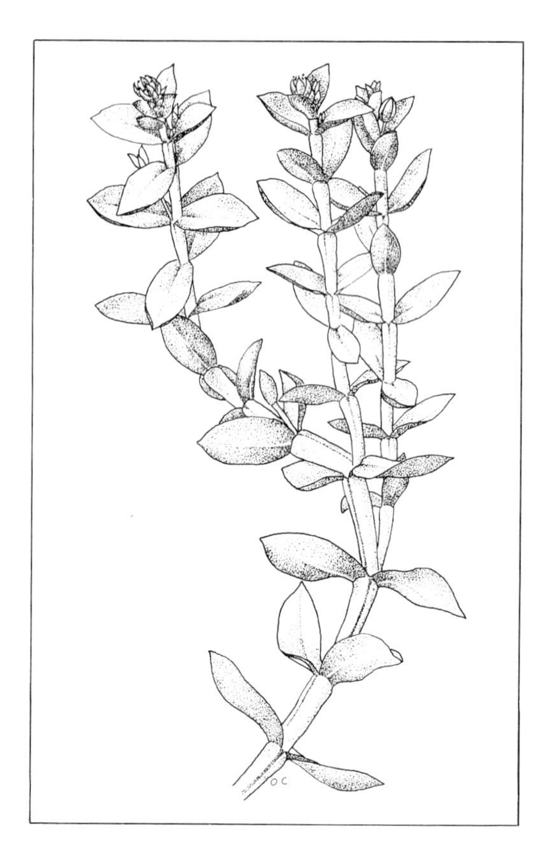

Other Names
Sea-purslane, sea-chickweed.

How to Recognize
This is a yellowish-green perennial with numerous, trailing, freely branching stems up to 30 cm high, starting as one plant, spreading, and rooting freely to form a dense mat of plants. The fleshy, smooth leaves vary in length from less than 1 cm to 3 cm and are narrow to broadly oval in shape. The greenish, inconspicuous flowers are borne singly in the axils of the upper branches or in several-flowered clusters. The fruiting capsules are globular and up to 8 mm long. This species is also known as *Honckenya peploides* (L.) Ehrh. Several different varieties have been defined on the basis of size and shape of leaves, and position of the flowers, among other characteristics.

Where to Find
Varieties of this plant grow on sandy beaches throughout Canada's coastline, as well as in northern Eurasia.

***Arenaria peploides* L.**
(Caryophyllaceae)

How to Use

The succulent young leaves of this plant have a pleasant, juicy texture and a rather strong flavour somewhat resembling that of cabbage. They can be eaten raw, alone or mixed with other greens as a salad, or lightly cooked with salt and butter. In Europe, the plants are pickled like samphire and glasswort (see p. 120) and are said to have a pleasant, pungent taste as a relish. The vitamin content of this plant is notable: a study by Hoffman *et al.* found the leaves to contain 42 mg vitamin C per 100 g fresh weight and 34 μg vitamin A per gram fresh weight in the sample analyzed.

Suggested Recipes

Seabeach-Sandwort Salad with Tuna

500 mL	fresh seabeach-sandwort shoots, washed and cut into small pieces	2 cups
	1 small can flaked tuna	
250 mL	celery, finely chopped	1 cup
125 mL	green onion, chopped	1/2 cup
125 mL	mayonnaise	1/2 cup
15 mL	lemon juice	1 tbsp
5 mL	soy sauce	1 tsp

Combine all ingredients in a salad bowl, mix gently, and chill. Serves 4.

Seabeach-Sandwort with Chopped Ham

1 L	fresh seabeach-sandwort shoots, cut into small pieces	4 cups
30 mL	green onion, chopped	2 tbsp
15 mL	vegetable oil	1 tbsp
	1 clove garlic, minced	
30 mL	water	2 tbsp
15 mL	fresh parsley, finely chopped	1 tbsp
500 mL	cooked lean ham, chopped	2 cups
	salt and pepper	

Wash and drain seabeach-sandwort shoots. In a skillet heat oil until slightly bubbly, then add onion and garlic. Stir about 15 seconds until browned, then add seabeach-sandwort, water, parsley, ham, and seasoning. Cook, stirring frequently, for 5 minutes, or until sandwort is tender but not mushy. Serve hot. Serves 4–6.

More for Your Interest

In Iceland, seabeach-sandwort plants are steeped in sour whey and allowed to ferment, then eaten as a sauerkraut. The Eskimos of Alaska picked the leaves in summer, placed them in a large pan, poured boiling water over them, and stored them in a barrel or sealskin poke until winter, when they were eaten with sugar.

Botanist Harold St. John, in "Sable Island, with a Catalogue of its Vascular Plants", noted that this plant was the choicest fodder for the "gangs" of wild ponies that abounded on Sable Island off Nova Scotia. These wild ponies are still found on Sable Island and are undoubtedly still enjoying seabeach-sandwort greens.

Strawberry Spinach

(Goosefoot Family)

Other Names

Indian strawberry, strawberry blite, Indian paint.

How to Recognize

Strawberry spinach is a smooth, erect annual with bright-green triangular to arrowhead-shaped leaves with wavy or coarsely toothed margins. It is closely related to the well-known, delicious weedy vegetable, lamb's quarters (*C. album* L.—see the first publication in this series, *Edible Garden Weeds of Canada*) and both of these plants are related to the common garden spinach. The flowers grow in tight, spherical clusters in the leaf axils and in spikes at the top of the stem and branches. In late summer and early fall they mature into brilliant masses of pulpy, scarlet fruits with the shape and texture of strawberries. This plant was formerly known as *Blitum capitatum* L.

Where to Find

Strawberry spinach occurs in all provinces of Canada except Prince Edward Island and Newfoundland, and is very abundant in parts of the North. It grows in a variety of habitats, from waste land and cultivated ground to clearings, burned areas, and gravelly river bars.

How to Use

In spring the tender young leaves and stems, and later the leaves by themselves, can be enjoyed raw or cooked as a nutritious green, rich in vitamins A and C. They can be substituted for spinach or lamb's quarters in any recipe, and many consider them superior to spinach in taste. They are good in salads, or lightly steamed and served with butter and lemon juice, or with breadcrumbs fried in butter as you might serve green beans. Because the plant is common in many remote areas, it is an excellent vegetable for hunters, campers, and hikers, and as such is a good survival food. The fruits also are edible, though rather bland, and can be used raw or cooked.

***Chenopodium capitatum* (L.) Aschers.**
(Chenopodiaceae)

Warning

Like its cultivated relative, spinach, strawberry spinach contains oxalate salts and, in some habitats, can have high concentrations of nitrates. Therefore, it should be eaten only in moderation, as one would eat spinach. A related species, *C. ambrosioides* L., does not commonly occur in Canada but is found in the southern United States and appears to be spreading northwards. It is ill-scented and toxic and should not be consumed.

Suggested Recipes

Strawberry Spinach Salad

500 mL	strawberry-spinach leaves	2 cups
125 mL	strawberry-spinach fruits	1/2 cup
30 mL	vinegar	2 tbsp
	6 slices bacon, cooked until crisp, with drippings	
	2 hard-boiled eggs, sliced	
	dash of garlic salt (optional)	
	dash of salt, pepper	

Wash, drain, and shred strawberry-spinach leaves and wash fruits. Place leaves and fruits in a bowl and sprinkle with vinegar. Crumble bacon into bits and add to salad with egg slices. Season with salt and pepper and garlic salt if desired. Just before serving, pour hot bacon drippings over and toss well. Serves 2.

Note: if vinegar is not available, try adding 125 mL (1/2 cup) of shredded leaves of sheep sorrel (*Rumex acetosella* L.), wood-sorrels (*Oxalis* species), *or* mountain sorrel (*Oxyria digyna*), see p. 138.

Baked Strawberry Spinach

1 L	fresh young strawberry-spinach leaves	4 cups
30 mL	olive oil	2 tbsp
	1 large onion, finely chopped	
	1 clove garlic, minced	
	2 eggs, well beaten	
125 mL	feta cheese, crumbled	1/2 cup
	salt and pepper	
125 mL	Parmesan cheese	1/2 cup
30 mL	butter *or* margarine	2 tbsp

Carefully wash strawberry spinach and pat dry with a towel. Heat olive oil in large skillet until it begins to bubble, then add onion and garlic and sauté until onion is transparent. Add strawberry spinach, cover tightly, and cook for 5 or 6 minutes. Remove from heat, and when greens have cooled slightly, stir in eggs and feta cheese, season to taste and pour into a medium-sized, buttered baking dish. Sprinkle Parmesan overtop and dot with butter *or* margarine. Bake at 190°C (375°F) about 10 minutes or until eggs set. Serve hot with garlic bread. Serves 3–4.

More for Your Interest

The bright-scarlet fruits of this plant often stain the shoes or feet of those walking through patches of the plant. Indian peoples, such as the Thompson and Carrier of British Columbia, crushed the fruits and used them to colour clothes, skins, basket materials, implements, and even the face and body. At first the stain is bright red, but with time it darkens to a purple or maroon shade.

Strawberry spinach was introduced to Europe from North America in the early seventeenth century and was grown in gardens for its ornamental qualities and for use as a food, but neither leaves nor fruits were highly valued for eating.

Glassworts

(Goosefoot Family)

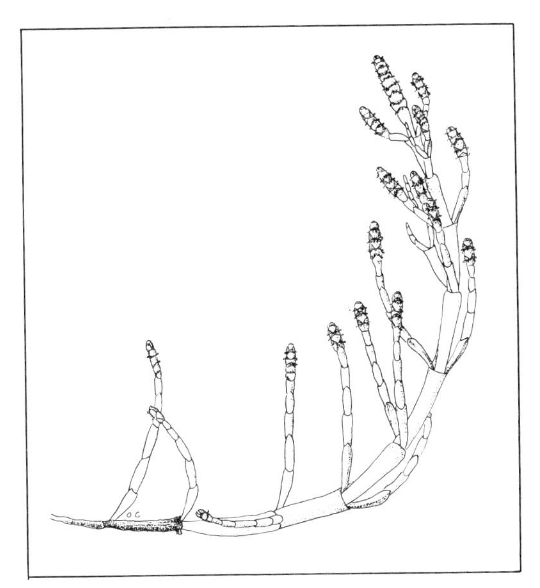

Salicornia virginica

Other Names
Pickleweed, false samphire, chicken-claws, pigeonfoot, leadgrass, bench asparagus.

How to Recognize
Glassworts are distinctive plants, once identified not easily forgotten. They are freely branching herbs, usually under 30 cm tall, with light-green or reddish, succulent stems 2 or 3 mm thick, smooth-skinned and jointed at the nodes. The leaves are minute and scale-like. The flowers are tiny and inconspicuous, borne in crowded spikes at the tips of short, erect branches.

Three glasswort species occur in Canada. One, the perennial, or woody, glasswort (*S. virginica* L.), is a matted perennial with slender rhizomes and long, prostrate, much-branched aerial stems bearing upright flowering stems. The other two, false samphire (*S. europaea* L.) and red glasswort (*S. rubra* A. Nels.), are short, erect annuals. The stems of the last two are often a reddish colour.

Where to Find
Perennial glasswort and false samphire grow along both Atlantic and Pacific coasts, on beaches, mudflats, and in saline marshes.

***Salicornia* species**
(Chenopodiaceae)

They are very tolerant of salt water and can withstand constant wetting with salt spray and occasional submersion from exceptionally high tides. Red glasswort grows in moist saline and alkaline soil from Manitoba to British Columbia, being especially common around the edges of alkaline lakes and ponds.

How to Use

The tender, succulent stem tips of the glassworts are pleasantly salty and can be eaten fresh right from the plants, used with dressing as a salad ingredient, or cooked lightly as a potherb. Perhaps the best use, however, and certainly the best known, is as a pickle or relish.

The shoots can be gathered from late spring to early fall, as long as care is taken to avoid the older, tougher stems. They should be thoroughly washed to remove excess salt; the juice of the stems is salty enough itself. If possible, the shoots should be soaked in cold, fresh water for an hour or more before being cooked or used fresh in salads. Of course, there is no need to add extra salt to the cooking water. In our experience, glasswort is not very good as a potherb, because unless it is cooked very lightly it becomes mushy and loses its flavour. However, when mixed with other potherbs or as an addition to soups and sauces it is quite palatable.

Suggested Recipes

Seaside Potato Salad

1 L	cold cooked potatoes, cubed	4 cups
500 mL	tender, young glasswort shoots, cut into bite-sized pieces	2 cups
50 mL	wine vinegar	1/4 cup
30 mL	grated onion	2 tbsp
125 mL	sour cream	1/2 cup
2 mL	paprika	1/2 tsp
30 mL	chopped parsley	2 tbsp
	2 hard-boiled eggs, sliced	
	4 large ripe olives, chopped	
	4 to 6 large, crisp lettuce leaves	

Mix together potatoes, glasswort, vinegar, onion, sour cream, paprika, and parsley. Chill and serve on lettuce leaves, garnished with sliced eggs and chopped olives. Serves 4–6.

Glasswort Omelette

125 mL	chopped onion	1/2 cup
50 mL	butter *or* margarine	1/4 cup
250 mL	tender, young glasswort shoots, cut into bite-sized pieces	1 cup
125 mL	milk	1/2 cup
	6 eggs, well beaten	
125 mL	sharp Canadian Cheddar cheese, grated	1/2 cup
	dash of Tabasco sauce	
	pepper and paprika to taste	

In a heavy skillet, sauté onion in butter *or* margarine until lightly browned. Combine other ingredients and add mixture to onion in skillet. Cook over medium heat. As bottom of omelette sets, loosen and lift with a spatula, tilting skillet to allow uncooked portion to run underneath; continue until omelette is almost dry on top. Lift omelette with spatula, fold or roll, and serve hot. Serves 3.

Glasswort–Sardine Salad

	1 small can sardines	
	1 hard-boiled egg	
	2 green onions, chopped	
125 mL	tender, young glasswort shoots, cut into bite-sized pieces	1/2 cup
125 mL	cooked green peas, chilled	1/2 cup
30 mL	mayonnaise	2 tbsp
	pepper and paprika to taste	
	1 lemon, cut in wedges	
	1 medium-sized avocado, peeled and sliced	

Chop and mix together sardines and hard-boiled eggs. Add onion, glasswort, peas, mayonnaise, and seasoning. Chill and serve garnished with lemon wedges and avocado slices. Serves 2–3.

Glasswort Pickles

3 L	tender, young glasswort shoots (approximately)	12 cups
1 L	wine- *or* cider-vinegar	4 cups
125 mL	sugar	1/2 cup
30 mL	pickling spices, mixed	2 tbsp
	2 or 3 bay leaves	
	2 medium-sized onions, sliced	
	1 chili pepper, chopped (optional)	

Wash glasswort thoroughly and cut into short lengths. Place in sterilized pickling jars (do not pack too firmly). Prepare pickling solution by boiling vinegar with other ingredients for 10 to 15 minutes. Pour hot solution over glasswort and seal jars. Let stand 4 to 6 weeks before using. You may wish to add salt if you like your pickles very briny, but keep in mind that glasswort is quite salty naturally. Makes about 12 medium-sized jars.

More for Your Interest

The name *Salicornia* is derived from the Latin *sal* for "salt" and *cornu* for "horn", alluding to the preferred saline habitat of the plant, and its horn-like branches. Charlotte Bringle Clarke, in her book *Edible and Useful Plants of California*, states that the term *glasswort* originates from the high soda content of the plants, which were said to be used in the past by glass and soap manufacturers. Others suggest that the name is simply derived from the translucent, glass-like appearance of the shoots, particularly after they are touched by frost.

Ducks and geese feed on the succulent stems of glasswort, especially in the fall when the seeds are borne in the mature branch tips.

In Europe, glasswort has been widely gathered in the past by country people, pickled and sold under the name *samphire*, but this term is more correctly applied to an entirely different seaside plant in the celery family (*Crithmum maritimum* L.), whose aromatic fleshy leaves were also prized for making pickles, although they are seldom used today. Glasswort has been used as a salad green, potherb, and pickling weed not only in Europe and North America but also in South Africa and the West Indies.

Stonecrops

(Orpine Family)

Other Names

Sedum, orpine; *S. roseum* is known as roseroot and is sometimes called "scurvy-grass", a name commonly used for plants containing vitamin C.

How to Recognize

There are at least eight species of *Sedum* native to Canada, and a number of others that have been introduced from Europe and elsewhere as ornamental plants for rock gardens and that are now found as garden escapes in many localities. Roseroot [*S. roseum* (L.) Scop., also known as *Rhodiola rosea* L.] is the most edible but all of the sedums can be eaten, and some, such as *S. divergens* Wats., are particularly deserving of attention.

In general the stonecrops are fleshy-leaved perennials (occasionally annuals or biennials) usually spreading by rhizomes or stolons and often forming dense patches. Most are under 10 cm high, short-stemmed, with leaves flattened or rounded in cross-section and often tightly crowded on vegetative shoots, though more widely spaced on flowering stems. The flowers, ranging in colour from white to yellow to rose-purple, are borne in flat-topped or rounded terminal clusters and are often very showy. The fruits are fleshy, erect, greenish to reddish follicles.

Roseroot grows from a thick, much-branched rootstock. The numerous leafy stems, up to 20 cm high, bear pale, elongated, somewhat spoon-shaped, smooth-edged or coarsely toothed leaves. The plants are dioecious, with male and female flowers on separate stems; the male flowers are bright yellow and the female ones usually purplish.

S. divergens has numerous, succulent, globular, berry-like leaves crowded in opposite pairs on short stems. The leaves are often reddish in colour, especially in exposed habitats. The flowers are bright yellow.

Where to Find

Most of the stonecrops grow in exposed rocky places or on sandy or gravelly soil. Being succulent, they can withstand periods of drought in the summer. Stonecrop species are to be found almost everywhere in Canada, although several, including *S. divergens*, *S. oreganum* Nutt., and *S. spathulifolium* Hook., occur only in British Columbia and the western United States. Roseroot is very common in parts of northern Canada, extending in range from the mountains of British Columbia and Alberta to Newfoundland.

***Sedum* species**
(Crassulaceae)

How to Use

All the stonecrops are edible, but the taste and texture vary with species. Some tend to be tough and some too acrid-tasting to be of any use, but the young leaves and shoots of most are just pleasantly tart and very juicy and succulent, especially in springtime. The shoots can be nibbled raw, mixed in salads, or lightly cooked and served with butter and salt. We have found that a touch of garlic salt enhances their flavour considerably. The stonecrops are a good survival food since they grow commonly along ocean cliffs, on rocky mountain tops, and, in the case of roseroot, in the Arctic tundra zone. They are usually very abundant where they do occur and are easy to harvest.

Samples of roseroot leaves were analyzed by Hoffman *et al.* and found to contain significant amounts of vitamin C (68 mg per 100 g fresh weight in one sample) and vitamin A (25 µg per g fresh weight).

Warning

Some of the *Sedum* species have emetic and cathartic properties and can cause headaches if consumed in large amounts. Eat only in moderation.

Suggested Recipes

Tuna Salad with Stonecrop

	1 can flaked tuna	
250 mL	young leaves and stem tips of stonecrop	1 cup
15 mL	capers	1 tbsp
50 mL	mayonnaise	1/4 cup
15 mL	soy sauce	1 tbsp
	salt and pepper	
	cold, crisp lettuce leaves	
	1 dozen cherry tomatoes	

Combine tuna, stonecrop, capers, mayonnaise, soy sauce, and seasoning. Line 3 salad bowls with lettuce leaves and spoon salad onto leaves. Garnish with cherry tomatoes. Serves 3.

Butter-Steamed Stonecrop *à la Polonaise*

250 mL	young leaves and stem tips of stonecrop	1 cup
125 mL	fine, dry breadcrumbs	1/2 cup
50 mL	butter	1/4 cup
	1 hard-boiled egg, chopped	
	salt and pepper	
	dash of garlic salt	
	chopped parsley for garnish	

Wash, trim ends, and remove old leaves from the stems of stonecrop (if stalks are too tough, use only leaves). Place in a saucepan with about 2 cm boiling, salted water, cover, and cook until tender (about 4 minutes). Drain and keep hot. Brown breadcrumbs in butter. Remove from heat and add chopped egg and seasonings. Sprinkle over hot stonecrop and serve garnished with parsley. Serves 2.

Stonecrop with Parmesan

250 mL	young leaves and stem tips of stonecrop	1 cup
30 mL	butter	2 tbsp
	Parmesan cheese, grated, to taste	
	dash of garlic salt (optional)	

Prepare stonecrop as in previous recipe. Boil in salted water until tender (about 4 minutes), then drain and place on broiling pan. Dot with butter, sprinkle generously with grated Parmesan, and garlic salt if desired. Broil until cheese melts, then serve hot on toast. Serves 2.

More for Your Interest

Roseroot (*S. roseum*) was eaten by the Eskimos of Alaska and the Inuit of northern Canada. The leaves and roots were placed in a sealskin poke with water, allowed to ferment, then frozen for winter use, when they would be served mixed with seal oil. Often roseroot shoots were combined with woolly lousewort shoots (*Pedicularis lanata*, see p. 151) or other edible greens.

The leaves of *S. divergens* were eaten by a number of coastal Indian groups of British Columbia who regarded them more as a type of berry than as a green. The Haida and Niska were particularly fond of them, and the former chewed them to freshen the mouth after eating such strong-tasting foods as fish grease. The Bella Coola Indians called them by the appropriately descriptive name of "strung salmon-roe", and the Kwakiutl called them "crow's strawberries".

Stonecrops are eaten by pikas, mountain beaver, and other small mammals. Various stonecrop species have been eaten in Europe and some are even cultivated in Holland and elsewhere for use in soups and salads. Many of the stonecrops make excellent rock-garden plants, being both attractive and easy to propagate. Some are apt to become weedy, however, and should be carefully controlled in a small garden. One of the common European species, *S. acre* L., occasionally found in Canadian gardens and as a garden escape, is often used in Europe for making wreaths for graves. Being succulent, it flowers readily, even when not rooted, and it soon roots itself and spreads over the grave, providing a permanent decoration.

A small parasitic flowering plant known as cancer-root (*Orobanche* species) is often found growing with the stonecrops. This leafless trumpet-shaped flower attaches itself to the stonecrop roots and derives its sustenance from this host. Many people think this is actually the stonecrop flower.

Fireweed

(Evening-Primrose Family)

Other Names
Blooming Sally, willowherb.

How to Recognize
One of Canada's most beautiful and widespread wildflowers, fireweed is a tall perennial herb, from 1 to 3 m, with numerous smooth-edged, lance-shaped leaves growing along the full length of the stem. Because the leaves closely resemble those of some of the narrow-leaved willows, the alternate name *willowherb* is sometimes used. By late spring a dense, elongated cluster of flower buds develops at the tip of the stem. The showy, 4-petalled flowers are pinkish to reddish-purple and mature from bottom to top of the cluster as the season progresses: thus, blooming can last for a month or so for any one plant. The long, narrow, fruiting capsules split open longitudinally when ripe to release numerous parachuted seeds that can be carried on the wind for many miles. Because of these far-travelling seeds, fireweed quickly establishes itself in new sites such as recently logged-off clearings or burns.

A related species, river beauty (*E. latifolium* L.), is also edible. It is shorter, up to 30 cm, and bushier than fireweed, with larger, brighter-pink flowers growing in smaller clusters.

Where to Find
Fireweed grows from coast to coast in Canada, and northwards to the Yukon and Northwest Territories. It also grows across Europe and the Soviet Union. It occurs in a wide variety of habitats from sea level to subalpine elevations but prefers open clearings, burned or logged-off sites, and moist meadows. Often it forms extensive patches, sometimes covering many acres and colouring entire hillsides purple. River beauty grows on river bars, gravelly banks, and moist scree slopes across northern Canada and southwards in the mountains.

***Epilobium angustifolium* L.**
(Onagraceae)

How to Use

The tender young shoots of the fireweed, harvested before the leaves have fully unfolded, are an excellent green vegetable either raw or cooked. One study, by Hoffman *et al.*, found fireweed leaves to contain up to 220 mg ascorbic acid (vitamin C) per 100 g fresh weight and 112 μg β-carotene (vitamin A) per gram fresh weight in the samples analyzed. These figures qualify fireweed as an exceptionally good source of these vitamins. River beauty was also found to be high in vitamins C and A.

Fireweed shoots should be gathered just at the stage when they snap off at the base like asparagus stalks. (The older, mature leaves and mature stalks can be eaten in an emergency, but are inclined to be bitter, and the stalks must be peeled first, as the outer part becomes tough and fibrous.) We have found that the juiciest, most tender shoots are from plants growing in moist, shady places. At lower elevations the best time to gather fireweed is in April and May, but in montane areas it can be harvested throughout the summer.

A good way to serve fireweed is to cut the young stems into small pieces, barely cover them with water, and boil them until soft, from 3 to 5 minutes, depending on the age and tenderness of the shoots. Serve hot with butter, as you would asparagus or green beans. You can also stir-fry the shoots in Oriental fashion by placing them in a hot skillet with about 15 mL (1 tbsp) vegetable oil and 30 mL (2 tbsp) water per 250 mL (1 cup) shoots, adding soy sauce to taste and cooking quickly until the greens have just softened but are still crisp inside.

Fireweed leaves are also well known as a tea substitute, as described in our book, *Wild Coffee and Tea Substitutes of Canada*. Fireweed flower buds are edible and make a colourful addition to salads and desserts.

Like many greens, fireweed has been found by some people to be slightly laxative, especially when first taken. Sometimes Indian people eat it as a spring tonic. We advise you to take only a small portion when you first try it.

Suggested Recipes

Cream of Fireweed Soup

250 mL	young fireweed shoots, washed and chopped	1 cup
250 mL	cold water	1 cup
30 mL	butter	2 tbsp
30 mL	flour	2 tbsp
250 mL	milk	1 cup
	salt and pepper	
	dash of garlic salt	
30 mL	sour cream (optional)	2 tbsp
15 mL	lemon juice (optional)	1 tbsp
	chopped fresh parsley	

Place fireweed and water in a saucepan, bring to a boil, reduce heat, and simmer 5 minutes. Drain, saving cooking water. Press fireweed through a sieve or purée in a blender and set aside. In a saucepan, melt butter, add flour, and blend until smooth. Add milk and cooking water from fireweed and heat, stirring gently, until mixture thickens. Add salt, pepper, garlic salt, and fireweed purée. Just before serving stir in sour cream and lemon juice if desired, and garnish with parsley. Serves 2.

Salad *Mélange*

250 mL	young fireweed shoots, washed and chopped	1 cup
250 mL	watercress sprigs	1 cup
	1 small head of lettuce, broken into bite-sized pieces	
	1 small cucumber, sliced	
	6 medium radishes, sliced	
	2 medium tomatoes, cut in chunks	
	2 medium carrots, grated or finely chopped	
	salt and pepper	
125 g	cream cheese	4 oz
15 mL	anchovy paste	1 tbsp
30 mL	cream	2 tbsp
125 mL	French dressing	½ cup
15 mL	dillweed, finely chopped	1 tbsp
	lemon slices	

Put fireweed and other vegetables into a large salad bowl, and sprinkle with salt and pepper. Blend together cream cheese, anchovy paste, cream, French dressing, and dillweed, and pour over vegetables. Toss thoroughly. Chill and serve garnished with lemon slices. Serves 4–6.

Chicken Summer Salad with Fireweed

500 mL	cooked chicken (*or* turkey), diced	2 cups
250 mL	young fireweed shoots, washed and chopped	1 cup
250 mL	diced celery	1 cup
	salt and pepper	
30 mL	mayonnaise *or* other salad dressing	2 tbsp
	3 hard-boiled eggs	
	3 tomatoes	
	a few ripe olives (optional)	

Mix chicken, fireweed, celery, salt, pepper, and mayonnaise *or* other dressing, chill and serve on crisp lettuce garnished with slices of hard-boiled egg, tomatoes, and ripe olives. Serves 4–6.

Salmon Loaf with Fireweed

500 g	cooked salmon	1 lb
500 mL	young fireweed shoots, washed and chopped	2 cups
250 mL	breadcrumbs	1 cup
	1 medium onion, finely chopped	
	2 stalks celery, finely chopped	
	2 eggs, well beaten	
15 mL	lemon juice	1 tbsp
30 mL	sour cream	2 tbsp
	salt and pepper	
50 mL	butter	1/4 cup
50 mL	flour	1/4 cup
250 mL	milk	1 cup
15 mL	chopped dillweed	1 tbsp

De-bone and flake salmon and mix in fireweed, breadcrumbs, onion, celery, eggs, lemon juice, sour cream, and seasoning. Place mixture in a buttered loaf tin and bake 45 to 60 minutes, or until firm, in a 160°C (325°F) oven. Meanwhile, prepare a white sauce by melting butter in a saucepan, blending in flour, adding milk, and stirring over moderate heat until sauce thickens. Season with salt and pepper and pour over baked salmon loaf. Garnish with dillweed. Serves 4.

Fireweed *au Gratin*

500 mL	young fireweed shoots, washed and chopped	2 cups
30 mL	butter	2 tbsp
30 mL	flour	2 tbsp
250 mL	milk	1 cup
	salt and pepper	
250 mL	breadcrumbs	1 cup
125 mL	sharp Cheddar cheese, grated	1/2 cup
1 mL	garlic salt	1/4 tsp

Barely cover fireweed with water, bring to a boil, reduce heat, and simmer about 3 minutes or until tender. Drain and place in buttered baking dish. In a saucepan or double boiler, melt butter and blend in flour. Add milk and heat gently, stirring frequently, until mixture thickens. Season with salt and pepper. Pour mixture over fireweed shoots. Mix together breadcrumbs, grated cheese, and garlic salt and sprinkle over sauce in baking dish. Bake 30 minutes at 160°C (325°F). Serves 2.

More for Your Interest

In the early trading days, French-Canadian voyageurs were very fond of fireweed as a potherb and called it *herbe frette*, or "cargo herb". On the Gaspé Peninsula of Quebec, the local inhabitants to this day call fireweed *asperge* (asparagus).

Fireweed stalks were a popular spring food for many Indian peoples. On the coast of British Columbia the young shoots were gathered in large quantities and eaten at feasts and family gatherings. The stems were split lengthwise with the thumbnail, spread open, dipped in oil or, within the last century, sprinkled with sugar or honey, and pulled through the teeth to separate the tender inner part from the fibrous outer skin. The outer part was either discarded, or dried and spun into twine for making fishnets. This practice was common among the Haida of the Queen Charlotte Islands.

The cottony seed fluff of fireweed was used by some Salish Indians of British Columbia and Washington State for spinning and weaving. It was mixed with other fibres, such as the wool of mountain goats or of small domesticated dogs bred on Vancouver Island and the lower mainland before the coming of Europeans. The mixture was then woven into blankets and clothing. The Saanich Indians of Vancouver Island stuffed mattresses with a blend of fireweed "cotton" and duck feathers.

In Alberta, the Blackfoot Indians rubbed fireweed flowers on mittens and rawhide thongs as waterproofing. The inner pith from the stems was dried, powdered, and rubbed on the hands and face in winter as a talc, to prevent chapping of the skin.

Fireweed's abundance and long blooming season make it especially valuable as a honey plant in Canada, and many beekeepers transport their hives each summer to areas where fireweed is plentiful. Mild-tasting, golden fireweed honey is sold in markets throughout Canada.

Seaside Plantain

(Plantain Family)

Other Names

Seashore plantain, rushlike plantain, goose-tongue.

How to Recognize

This species is one of several edible plantains to be found in Canada. Two weedy species, broad-leaved plantain (*P. major* L.) and narrow-leaved plantain (*P. lanceolata* L.), are discussed in *Edible Garden Weeds of Canada* (see table, p. 25). The seaside plantain is a perennial herb growing from 1 to a few long, thick roots. Numerous narrow, pointed leaves, up to 20 cm or more in length, rise from a single or short-branched crown. The leaves are smooth and slightly fleshy. One to several flowering stems, usually slightly longer than the leaves, are borne in late spring or early summer. The small, greenish flowers, crowded in dense terminal spikes, ripen into brown, ovoid capsules, each containing 3 to 4 seeds. This species is also known as *P. juncoides* Lam.

Where to Find

Seaside plantain grows along sandy or gravelly marine shorelines, in salt marshes, and in crevices of large rocks along both Pacific and Atlantic coastlines and along the shores of Hudson, James, and Ungava bays.

How to Use

The succulent leaves of seaside plantain have a pleasant salty flavour and can be used as a salad vegetable when young and tender. The older tougher leaves usually require cooking, and are quite good when washed, cut in pieces, boiled, and served like string beans. They can also be cooked in soups and stews and are an ideal ingredient in seaside camp chowders. The leaves can be pickled in spiced vinegar, like glasswort and samphire (see *Salicornia* species, p. 117).

***Plantago maritima* L.**
(Plantaginaceae)

seaside plantain
(*Plantago maritima*)

arrow-grass (*Triglochin maritimum*)
Poisonous: do not eat
see Warning

Warning

A similar looking, though botanically unrelated, species, arrow-grass (*Triglochin maritimum* L.), is sometimes confused with seaside plantain. It is a grass-like perennial with numerous, erect, succulent, sheathing leaves growing in a basal cluster and with long spikes of small greenish flowers on stems up to 1 m tall, considerably longer than the leaves. This plant grows in tidal flats, salt marshes, and on muddy shorelines. The fleshy white leaf bases, harvested when young, are edible and sweet, but the mature green leaves and flower stalks are toxic and have caused poisoning of livestock.

Suggested Recipes

Plantain Seaside Salad

250 mL	fresh-cooked *or* canned abalone	1 cup
500 mL	young plantain leaves, cut in bite-sized pieces	2 cups
250 mL	fresh-cooked *or* canned crabmeat *or* shrimp	1 cup
250 mL	sour cream	1 cup
30 mL	onion, chopped	2 tbsp
	4 to 6 radishes, thinly sliced	
125 mL	celery, finely chopped	1/2 cup
30 mL	olive oil	2 tbsp
	juice of 1 lemon	
	salt and pepper	
	6 large lettuce leaves	
	2 hard-boiled eggs, sliced	

Mix together all but the last 2 ingredients. Chill well and serve on lettuce leaves in individual dishes. Garnish with egg slices. Serves 6.

Seaside Plantain Greens

500 mL	young plantain leaves, cut in slivers	2 cups
50 mL	breadcrumbs	1/4 cup
30 mL	butter *or* margarine	2 tbsp
125 mL	grated Cheddar cheese	1/2 cup

Place plantain in a saucepan with a small amount of boiling water, cover, and simmer for no longer than 10 minutes. Do not let plantain become mushy by boiling too long. Drain and serve with hot breadcrumbs fried in butter, topped with grated cheese. Serves 2–3.

Seaside Plantain–Salmon Burgers

500 g	canned *or* fresh cooked salmon	1 lb
375 mL	young plantain leaves, finely chopped	1 1/2 cups
30 mL	raw onion, finely chopped	2 tbsp
30 mL	celery, finely chopped	2 tbsp
30 mL	mayonnaise	2 tbsp
	a few drops Tabasco sauce	
	salt and pepper	
	6 hamburger buns	
	butter for spreading	
	6 slices Cheddar cheese	
	12 bacon strips, fried crisp	
	tomato slices (optional)	

Put salmon into a mixing bowl (with the liquid if you use canned salmon). Add plantain, onion, celery, mayonnaise, Tabasco, and seasoning. Mix well. Open buns, spread each half with butter, and toast flat on grill. Spread each half with salmon mixture, top with cheese, and broil until cheese melts (about 3 minutes). Before serving, top each half with a slice of bacon and tomato slices if desired. Serves 6.

More for Your Interest
Fishermen of Nova Scotia and New England use seaside plantain extensively as a vegetable. They call it by the name of *goose-tongue*, apparently alluding to the long, narrow shape of the leaves. European sailors, especially the French, used to boil the leaves in a broth on their sea voyages and also ate them as a salad green. In Wales the plants have been cultivated as sheep fodder.

Arrow-grass, which closely resembles seaside plantain (see Warning above), is also often called goose-tongue. The Coast Salish Indians of British Columbia relished the whitish leaf bases as a spring vegetable: they have a mild, sweet taste, reminiscent of cucumber. Within living memory the Haida Indians canned them as a winter food. The green leaves and mature plants are toxic, however, and should never be eaten.

Mountain Sorrel

(Knotweed Family)

Other Names
Sometimes called "scurvy-grass", a name given to several different greens having a high vitamin C content.

How to Recognize
This is a perennial herb growing from a short rootstock. The long-stemmed leaves, up to 8 cm in length, are clustered at the base and have rounded, bright-green, somewhat succulent blades up to 5 cm in width. The small, reddish-green to crimson flowers are borne in dense, elongated clusters on erect stems up to 30 cm high. The fruits are thin and flat, each surrounded by a broad, transparent wing.

Where to Find
Mountain sorrel is widely distributed in the Canadian Arctic and in alpine regions further south. It is common in cool, moist ravines and on slopes where the snow remains late. The growth can be very luxuriant below bird nesting sites and near human habitations where the soil is rich.

Oxyria digyna **(L.) Hill**
(Polygonaceae)

How to Use

The succulent leaves, like those of sheep sorrel (*Rumex acetosella* L.) and wood-sorrels, an *Oxalis* species, (see table, p. 25) are pleasantly acid in flavour and can be eaten raw in salads, cooked as a potherb, or stewed like rhubarb. A study by Hoffman *et al*. found a sample of the leaves and stems to contain 40 mg vitamin C per 100 g fresh weight and 53 μg vitamin A per gram fresh weight. Because of its high vitamin content, wide distribution, and local abundance, it is considered by A.E. Porsild, one of the foremost botanists in the study of edible plants of the Arctic, to be among the most important wild vegetables of the North. It is excellent mixed raw with other wild greens, such as watercress [*Rorippa nasturtium-aquaticum* (L.) Schinz. & Thell.] and other plants in the mustard family, and is very good as a garnish for salads of lettuce and other garden vegetables. When gathered young, before flowering time, mountain sorrel compares favourably with the cultivated French sorrel (*Rumex acetosa* L.). It is especially delicious served with fish or rice or cooked as a thick soup or purée.

Warning

Mountain sorrel, like sheep sorrel, wood-sorrels, and rhubarb, contains oxalic acid, the compound that gives these plants their pleasantly sour taste. However, since oxalic acid can be poisonous if taken in excess, it is best to eat mountain sorrel only in moderation. (Remember also that although the leaves of the sorrels are edible, those of rhubarb are not; eat only the stalks of rhubarb, never the leaves.)

Suggested Recipes

Mountain Sorrel Glacier Salad

500 mL	young mountain sorrel leaves, shredded	2 cups
500 mL	white cabbage, finely sliced	2 cups
125 mL	sour cream	½ cup
250 mL	small curd cottage cheese	1 cup
125 mL	garlic sausage, finely chopped	½ cup
	2 medium green onions, finely chopped	
	salt and pepper	
	6 radishes, thinly sliced	

Mix together mountain sorrel, cabbage, and sour cream in a salad bowl. Blend cottage cheese, sausage, onion, and seasoning together and spread over the cabbage–sorrel mixture. Garnish with sliced radishes and serve well chilled. Serves 4.

Mountain Sorrel Soup

1 L	young stems and leaves of mountain sorrel	4 cups
30 mL	flour	2 tbsp
250 mL	sour cream	1 cup
	salt and pepper	
	half a lemon, sliced	

Place sorrel in a saucepan, barely cover with cold water and bring to a boil. Cook only 1 to 2 minutes, until the leaves become soft and tender, drain, then press through a sieve or purée in blender and return to saucepan. In a cup, blend flour with a little cold water until it forms a smooth soupy paste. Add this gradually to the puréed sorrel, mixing well to avoid lumps. Gradually add sour cream, mixing thoroughly until the cream is well blended. Season and heat to just below boiling point and serve garnished with lemon slices. Serves 4.

Mountain Sorrel–Salmon Supper

500 g	cold cooked *or* drained canned salmon	1 lb
	4 fresh large mushrooms, sliced	
500 mL	mountain sorrel leaves	2 cups
	1 medium cucumber, thinly sliced	
75 mL	olive oil	1/3 cup
30 mL	white vinegar	2 tbsp
1 mL	dry mustard	1/4 tsp
	salt and pepper	
	2 hard-cooked egg yolks	
125 mL	mayonnaise	1/2 cup
125 mL	sour cream	1/2 cup
30 mL	chopped parsley	2 tbsp
15 mL	chopped chives	1 tbsp
30 mL	capers (optional)	2 tbsp

Flake and de-bone salmon and place in a salad dish. Cover with sliced mushrooms and arrange the sorrel and cucumbers around the salmon. In a separate dish, mix olive oil, vinegar, mustard, and seasoning and pour over the salad. Serve with a dressing of egg yolks blended with mayonnaise, sour cream, parsley, chives, and capers if desired. Serves 3–4.

More for Your Interest

The leaves and stems of mountain sorrel are consumed in quantity by caribou, muskoxen, and geese. The fleshy rhizomes are eaten by Arctic hares and lemmings.

The Inuit prize this plant as a vegetable and eat it both fresh and preserved in seal oil. They often boil the leaves together with those of other Arctic plants, such as some of the peppery cresses, then place them with water in a sealskin poke and allow them to ferment. The fermented mixture is then stored frozen in underground caches. For use, it is thawed slightly, chopped up with seal oil, and served as a kind of "Eskimo ice cream", often as a dressing for meat.

Mountain Bistorts

(Knotweed Family)

Other Names
Mountain knotweed, alpine bistort, snakeweed; *P. viviparum* is known as serpent grass, and *P. bistortoides* as American bistort.

How to Recognize
Mountain bistorts are perennials, growing from short, thick, fleshy rhizomes. The leaves, mostly basal, are long-stalked and elliptic to lance-shaped, tapering at the base. The flowering stems are simple and erect, usually bearing 1 or 2 strongly reduced, somewhat triangular leaves. Above each leaf the stem is encircled with a papery sheath. The small, pinkish to white flowers are borne in a single, terminal, compact cluster. *P. viviparum,* which grows about 30 cm high, is quite variable in size and leaf shape; except in var. *macounii* (Small) Hult., it is generally smaller and less robust than *P. bistortoides* which grows to 60 cm in height. Some of the lower flowers of *P. viviparum* are replaced by small pinkish bulblets that drop to the ground when mature and grow into new plants. Another related species, also edible, is *P. bistorta* L. Its leaves are broadly rounded at the base and the upper leafstalks are winged.

***Polygonum viviparum* L. and**
***P. bistortoides* Pursh**
(Polygonaceae)

Polygonum viviparum

Mountain bistorts can be distinguished from the numerous other *Polygonum* species to be found in Canada by their short, thick rhizomes, mainly basal leaves, and flowers borne in solitary terminal clusters on erect stems.

Where to Find

These plants grow, usually at higher elevations, in shaded woods, moist meadows, along streambanks, and on alpine slopes. *P. viviparum* is abundant in Arctic regions and in montane areas from Newfoundland to British Columbia; *P. bistortoides* occurs in southern British Columbia and southwestern Alberta; and *P. bistorta* grows in the grassy tundra of the North, in the Yukon, and District of Mackenzie, as well as in Alaska.

How to Use

These species are better known for their delicious, nut-like rootstocks than for their edible greens. However, the greens are also quite good, especially when young, and make a nutritious addition to any wild food diet. (A study by Hoffman *et al.* found *P. viviparum* leaves to contain nearly 160 mg ascorbic acid per 100 g fresh weight in the sample analyzed.) In some localities the leaves, like the rootstocks, can be slightly astringent and are sometimes improved with cooking, although we found both rootstocks and leaves very palatable raw. If you do wish to cook them, they can simply be steamed and eaten with butter and salt, or used as an ingredient in soups, stews, and casseroles. The leaf stems should be removed before eating as they tend to be tough. The small purplish bulblets of *P. viviparum*, which replace the lower flowers in the cluster, can be stripped from the stem to make a sweet nibble or a pleasant addition to salads.

Warning

These greens and those of other knotweeds contain hydrocyanic acid and should be eaten in moderation. There is also a possibility that some species can cause photosensitization (hypersensitivity to light) in some individuals. However, this would occur only after eating large quantities and then being exposed to intense sunlight for a long period of time.

Suggested Recipes

Sautéed Bistort Greens

500 mL	young bistort leaf blades	2 cups
50 mL	butter *or* margarine	1/4 cup
	onion salt	
	pepper	
	6 slices cooked, crumbled bacon	

Wash and remove stems from bistort leaves. In a heavy frying pan, melt butter *or* margarine and cook bistort leaf blades under cover until tender (5 to 6 minutes). Season with onion salt and pepper and garnish with bacon. Serve for breakfast with eggs and toast. Serves 2.

Bistort–Potato Salad

	5 medium-sized potatoes	
	1 medium-sized mild red onion	
	10 slices cooked, crumbled bacon, with drippings	
500 mL	young bistort leaf blades	2 cups
125 mL	celery stalks, thinly sliced	1/2 cup
30 mL	salad oil	2 tbsp
125 mL	sour cream	1/2 cup
30 mL	lemon juice	2 tbsp
30 mL	grated Parmesan cheese	2 tbsp
	dash of paprika	
	salt and pepper	

Boil potatoes until tender, then drain, cool, peel, and cut into 1 cm (1/2 in.) cubes. Thinly slice onion, separate into rings, and add to potatoes, along with crumbled bacon. Add bistort leaves to hot bacon drippings and sauté over medium heat for about 5 minutes, stirring from time to time, until tender. Drain, cool, and add to potato mixture. Add remaining ingredients, mix well, and serve cold. Serves 4.

Bistort with Garlic Sausage

30 mL	butter *or* margarine (first amount)	2 tbsp
1 L	young bistort leaf blades	4 cups
50 mL	water	1/4 cup
500 g	garlic sausage, chopped	1 lb
30 mL	butter *or* margarine (second amount)	2 tbsp
	1 medium onion, sliced (optional)	

In frying pan melt butter *or* margarine (first amount). Add bistort leaves and water and cook under cover about 5 minutes, until leaves are tender. Fry garlic sausage separately in butter *or* margarine (second amount), adding onion if desired. Combine sausage with bistort leaves and serve hot. Serves 3–4.

More for Your Interest

In northern Canada, Alaska, and north-eastern Siberia the native peoples eat the roots of *P. viviparum* as we would eat nuts or raisins. They gather and eat them raw or roast them in the ashes of a fire. Some say they are almond-like in flavour, others claim they are not unlike small potatoes.

The roots of *P. bistorta* are also eaten; in times of scarcity in Russia and Siberia they are roasted as a substitute for bread. In the southern counties of England, the young shoots of *P. bistorta* were formerly much used as an ingredient in herb puddings and as a green vegetable. The herb puddings were traditionally made at Eastertime and were known as Easter Ledger puddings. They were made by mixing together various combinations of bistort leaves, young nettle tops, dandelion leaves, and other wild greens, which were then cooked, strained, chopped, and added to beaten egg, hard-boiled chopped egg, butter, salt, and pepper. The mixture was heated through, then transferred to a hot pudding basin to shape. The puddings were usually eaten with veal. The tradition of making Easter pudding with this bistort, locally called "dock", was revived in 1971, with the announcement of the first World Championship Dock Pudding Contest, held at Calder Valley in Yorkshire. According to Richard Mabey in *Food for Free*, there were over fifty competitors from Calder Valley alone.

Thimbleberry and Salmonberry

(Rose Family)

How to Recognize

Thimbleberry (*Rubus parviflorus* Nutt.) and salmonberry (*R. spectabilis* Pursh) are relatives of the raspberry and blackberry, and are well known as edible wild fruits. Few people realize that their green sprouts are also edible.

Both are erect, bushy shrubs. Thimbleberry grows up to 2.5 m tall, with unarmed stems and thin, shredded bark. The long-stalked leaves are large, simple, and maple-leaf-like, with 3 to 7 pointed lobes and finely toothed margins. The flowers, white or pinkish-tinged, are large and borne in few-flowered terminal clusters. The bright-red fruits are soft and fleshy, with many small drupelets forming a shallow cup, or "thimble".

Salmonberry may exceed 4 m in height. Its freely branching stems are usually armed with small, sharp prickles. The leaves, like those of raspberry, are compound, with 2 lateral leaflets and 1 larger terminal leaflet. The pink to reddish-purple flowers, which open early in spring often before the leaves expand, are usually solitary, and borne on short branches along the stems. The large, watery fruits range in colour from golden yellow to ruby to almost black.

Rubus parviflorus (left)
Rubus spectabilis (right)

Rubus parviflorus **Nutt. and**
R. spectabilis **Pursh**
(Rosaceae)

A number of other bushy *Rubus* species have edible sprouts, as do their relatives, the wild roses (*Rosa* species). However, thimbleberry and salmonberry sprouts are among the best and are given here as examples.

Where to Find

Thimbleberry grows, often in dense thickets, along roadsides and in open woods and clearings from British Columbia to the Great Lakes region of Ontario. Salmonberry is a Pacific coastal species, and is one of the major understory shrubs in damp coniferous forests, in swamps, and along streambanks west of the Coast and Cascade mountains in British Columbia.

How to Use

The succulent new-growth sucker shoots of both these species, and some other *Rubus* species as well, can be broken off at ground level, peeled, and eaten in spring, either raw or lightly cooked. They are sweet and juicy only when tender enough to snap off with the fingers, like asparagus shoots; as soon as they become woody they are no longer edible. They are good as vegetable sticks, for dipping in cream cheese or any kind of dip. When cut in pieces and lightly steamed in a small amount of water, they can be served as an asparagus, seasoned with butter, salt, and lemon juice, or cheese sauce. They can also be cooked in soups, stews, and casseroles.

Remember to be selective in your harvesting of these shoots: if all the shoots are picked from a plant, it will not be able to regenerate itself.

Warning
Do not eat the young shoots and twigs of plum, cherry, apple, peach, and pear (all of which are members of the rose family), because the foliage of these species contains cyanogenic glycosides, and is known to be a frequent cause of livestock poisoning.

Suggested Recipes
Salmonberry sprouts can be substituted for thimbleberry sprouts in the following recipes.

Thimbleberry Sprouts with Cheese

30 mL	butter *or* margarine	2 tbsp
	1 medium onion, minced	
500 mL	young thimbleberry sprouts, peeled and cut into short segments	2 cups
	salt and pepper	
125 mL	sharp Cheddar cheese, grated	1/2 cup

Melt butter *or* margarine in a frying pan, add onion, and cook over medium heat until onion is soft (about 5 minutes). Add sprouts and seasoning, and cook an additional 5 minutes, or until sprouts are tender. Add cheese and continue cooking until cheese is melted. Serve hot on toast. Serves 4.

Thimbleberry Sprout Salad with Shrimps

250 mL	young thimbleberry sprouts, peeled and cut into short segments	1 cup
	2 large stalks celery, cut in thin, slanting pieces	
125 mL	fresh, cooked shrimps	1/2 cup
15 mL	salad oil	1 tbsp
15 mL	soy sauce	1 tbsp
15 mL	wine vinegar	1 tbsp
	salt and pepper	

Put sprouts, celery, and shrimps into a salad bowl. Add salad oil, soy sauce, vinegar, and seasoning, mix well and chill. Serve with cold meat or smoked salmon. Serves 4.

More for Your Interest

Salmonberry and thimbleberry shoots were a popular spring vegetable among the Indians of coastal British Columbia, especially with the children. Early travellers to the west coast of Vancouver Island recounted seeing entire canoes laden with the shoots, being taken back to the villages where they would be consumed at large feasts. Usually they were eaten raw, with plenty of oil or fish grease and, in later times, were dipped in sugar. The fruits were also valued as a food (see our publication, *Edible Wild Fruits and Nuts of Canada*), and the leaves were sometimes made into tea (see our publication, *Wild Coffee and Tea Substitutes of Canada*).

Woolly Lousewort

(Figwort Family)

Other Names
Woolly fernwort, woolly fernweed, woolly pedicularis, "bumblebee plant".

How to Recognize
This is a low perennial growing 10 to 25 cm tall from a branching, well-developed, lemon-yellow taproot. The numerous, small leaves are finely divided like those of a fern, and are crowded along a short, simple stem. The flowers are small, rose-coloured, and tubular, borne in a dense, oblong, terminal cluster. The upper stem and flower cluster are densely woolly-white, a good identification feature of this plant. This species is also known as *P. kanei* E. Durand. Other related species, also edible, include hairy lousewort (*P. hirsuta* L.), Arctic lousewort (*P. arctica* R. Br.), and *P. sudetica* Willd.

Where to Find
Locally abundant in the North, this species grows on dry, stony tundra, mostly above the Arctic Circle, extending southwards in the high alpine regions of the mountains. Hairy and Arctic louseworts and *P. sudetica* are also found in the Arctic tundra zone, and Arctic lousewort and *P. sudetica* also occur in alpine habitats further south.

How to Use
Woolly lousewort, though locally abundant in some areas, is quite rare in others. It is a beautiful plant and, as harvesting it generally results in the destruction of the entire plant, it should be used only with the greatest discretion. It should be regarded as a survival food except where it is very common or where its destruction is inevitable because of land clearing or development.

The roots and tender young stems and leaves of the woolly lousewort and its relatives are edible either raw or simmered in a small quantity of water until tender. They are among the few wild vegetables to be found in the Far North, where the plants are very common in some localities. The shoots are best for eating before flowering time, early in the summer. The woolly texture of *P. lanata* greens may bother some people, but the other species mentioned are not as fuzzy. The roots of all these species can be prepared and eaten like carrots.

***Pedicularis lanata* Cham. & Schlecht.**
(Scrophulariaceae)

Suggested Recipes

Soured Lousewort, Eskimo Style

Gather and wash shoots from young lousewort plants. Put them into a watertight sealskin poke (bag), or a large bucket or plastic bag, cover with water, and allow greens to ferment for the rest of the summer, until just before the ground freezes. Place in an underground pit and store frozen for winter. To use, cut the frozen lousewort into pieces and allow to thaw partially. Add seal oil and sugar to taste, mix thoroughly, and serve cold, as a kind of "Eskimo ice cream". (Souring, or fermenting, is a common method of preparing greens among the Inuit.)

Steamed Lousewort

In early summer dig lousewort roots, leaving tops (stems and leaves) attached. Clean and wash thoroughly. Place the whole plants in a saucepan with a small amount of salted water, bring to a boil, reduce heat, and simmer about 10 minutes, or until roots are tender. Drain and serve hot with butter and lemon juice or garnish with grated Cheddar cheese and serve with eggs.

More for Your Interest

Inuit children like to suck the sweet nectar from the base of the long, tubular flowers of the woolly and Arctic louseworts. In parts of the U.S.S.R., the leaves of woolly lousewort are used to make tea. The Eskimos of Alaska mix the shoots with the leaves of roseroot (*Sedum roseum*, see p. 121) and other greens to store for winter. To this day, in northern Canada, Inuit people dig the roots of the woolly lousewort as a vegetable. The tops are browsed by caribou and reindeer.

Many, if not all, the louseworts are semi-parasitic, partially deriving their nourishment from the roots of other plants. For this reason they are difficult to transplant or grow in gardens. The Latin name *Pedicularis,* meaning "pertaining to lice", was given to the plant because it was believed that livestock eating louseworts would be more susceptible to infestation with lice.

Common Veronica

(Figwort Family)

Other Names
American brooklime, water speedwell.

How to Recognize
A perennial growing from shallow, creeping rhizomes, the common veronica has simple, somewhat weak, succulent stems up to 1 m long. The leaves grow in opposite pairs and are lance-shaped to elliptic with smooth or evenly toothed edges. The small, blue flowers are borne in open clusters at the leaf axils towards the ends of the stems. The 4-parted corolla drops off easily from the stalk when mature. The fruiting capsules are fleshy and green and slightly notched at the tips. There are a number of other *Veronica* species to be found in various parts of Canada, all edible.

Veronica americana **Schwein. ex Benth.**
(Scrophulariaceae)

Where to Find
The common veronica grows in damp places—in marshes and along streams and lake-margins—from the lowlands to moderate elevations in the mountains. It is widespread in Canada and all of temperate North America.

How to Use
The young leaves and succulent stems of this veronica, and other species as well, are slightly peppery or sharp-tasting, and are good eaten raw in salads and sandwiches or cooked as a potherb. It is a good mixer with other greens, especially watercress, which is often found growing with veronica. As early as the eighteenth century, veronica was used as an antiscorbutic, said to surpass even watercress in vitamin C content. The leaves are recommended by some for making a vitamin-rich tea.

In the North, veronica is particularly abundant in streams around the abandoned dwellings of early prospectors and trappers. Along with watercress, it is one of the most common and most abundant salad vegetables and potherbs of streams and ditches, and is valuable because of its long growing season.

Warning
Do not collect veronica from streams or ditches in populated areas, where the water may be polluted. When in doubt, add a drop of bleach to the water in which you are washing the greens, or use a halazone tablet that releases purifying chlorine gas, available from most drugstores.

Suggested Recipes

Arctic Char–Veronica Sandwich Spread

125 mL	leaves and young stems of veronica, finely shredded	1/2 cup
85 g	Arctic char spread	3 oz
30 mL	mayonnaise	2 tbsp
	prepared horseradish, to taste	
	salt and pepper	
	dash of onion powder	

Combine all ingredients, mix well, and spread on brown bread or toast. Makes enough for 4 sandwiches.

Veronica–Raw Mushroom Salad

250 mL	leaves and young stems of veronica, shredded	1 cup
250 mL	celery stalks, chopped	1 cup
15 mL	green onion, finely chopped	1 tbsp
	half a clove of garlic, minced	
50 mL	sour cream	1/4 cup
15 mL	olive oil	1 tbsp
	juice of half a lemon	
	salt and pepper	
	6 large, firm mushrooms, sliced	
	crisp lettuce leaves	
15 mL	chopped parsley	1 tbsp

Mix together veronica, celery, onion, garlic, sour cream, olive oil, lemon juice, and seasoning. Add mushrooms and stir until well blended. Chill and serve in individual salad bowls lined with lettuce. Garnish with chopped parsley. Serves 4.

Veronica–Watercress Trapper Sandwich Spread

125 mL	young leaves of veronica, finely shredded	1/2 cup
125 mL	young shoots of watercress	1/2 cup
30 mL	sour cream	2 tbsp
	small package of soft cream cheese	
	1 hard-boiled egg, finely chopped	
	salt, pepper, and paprika	

Combine veronica and watercress with sour cream, cream cheese, and egg, and mix well until fluffy. Add seasonings and spread on brown or rye bread. Makes enough for 4 sandwiches.

More for Your Interest

The genus name of this plant, according to Lewis J. Clark in *Wild Flowers of British Columbia*, is derived from Saint Veronica, canonized because she had wiped the face of Christ as he toiled with the cross towards Calvary Hill. Her kerchief then bore the *vera iconica*, or "true likeness". (However Frère Marie-Victorin, in his *Flore laurentienne* doubts that the name is derived from the saint.) In medieval times people linked Saint Veronica's name with this modest little country flower and invested it with miraculous healing powers. For centuries herbalists have used *Veronica* species for treating a host of ailments, particularly for urinary and kidney complaints and as a blood purifier.

The common name *brooklime* is an early English term and refers to the habitat of the plant; it grows in soft wet mud along brooks, where birds may become trapped, or "limed", an expression for ensnaring birds with sticky materials. The name *speedwell* is an old English blessing to departing guests, equivalent to "God speed" or "God bless you".

Wild Violets

(Violet Family)

How to Recognize
There are over thirty wild violet species in Canada. Most are easy to recognize, since they closely resemble our cultivated violets. They are low-growing annual or perennial herbs with leaves growing directly from the rootstock or from runners. The leaves are usually heart-shaped or rounded and basally lobed, although in some species the leaves are elongated or finely divided. The leaf edges are usually toothed, or serrated. The flowers are generally of the typical pansy form, with 2 upper petals, 2 lateral ones and 1, usually larger, lower one. Flower colour ranges from purple to mauve to white to yellow, and often the veins at the center of the petals are dark coloured. The lower petal is modified at the base to form a small sac or elongated spur. The fruits are rounded capsules that pop open when ripe, often throwing the seeds a considerable distance. Many violets also produce small, inconspicuous flowers at ground level, which are capable of developing fertile seed without even opening.

***Viola* species**
(Violaceae)

Viola canadensis

Among the most widespread species in Canada are the highly variable Canada violet (*Viola canadensis* L.), with white petals, often purplish-tinged on the back and yellow at the base, and the blue, or hooked, violet (*V. adunca* Sm.), with deep violet or violet-blue flowers. Occasionally the scented English, or sweet violet (*V. odorata* L.), with violet or white flowers, is found in woods as a garden escape.

Where to Find
Wild violets are found throughout Canada, except in the extreme North, in moist woods to open, dry, sandy areas. Canada violet occurs from southwestern Quebec to British Columbia. (Some botanists refer to the western variety of this violet as *V. rugulosa* Greene.) Varieties of the hooked violet are found from Newfoundland to British Columbia.

How to Use
The leaves of all the violet species are mild and quite edible, especially when young and tender. They can be eaten raw as a nibble or salad green, or cooked alone or in soups, stews, and casseroles. They are exceptionally high in vitamins C and A. A nutritious though somewhat bland tea can be made from the leaves simply by steeping a generous handful in a medium-sized teapot for about 5 minutes.

Violet flowers also are edible. Those of the scented English violet, found in many Canadian gardens, are especially good and can be used to make a wide variety of candies and conserves. Any of the violet flowers make a colourful and novel garnish for salads and gelatin desserts and can be eaten by themselves or along with the leaves.

Warning
According to Lewis and Elvin-Lewis in *Medical Botany*, the rhizomes and seeds of the scented violet (*V. odorata*) can be harmful if eaten; therefore do not eat the seeds, fruits, or rhizomes of any violets.

Suggested Recipes

Violet Greens

500 mL	fresh violet leaves	2 cups
10 mL	butter	2 tsp
	salt	
10 mL	lemon juice	2 tsp

Place violet leaves in a small saucepan, preferably in a vegetable steamer. Use a minimum of water (about 1 cm in the bottom of the pan should be sufficient). Cover, bring to a boil, then reduce heat and allow to steam 5 minutes. Meanwhile, melt butter and add salt and lemon juice. Drain cooked leaves, pour the butter mixture over them, and serve immediately. Serves 2.

Violet and Watercress Salad with Sour Cream

	1 clove garlic	
250 mL	fresh violet leaves	1 cup
250 mL	watercress, chopped	1 cup
	1 medium-sized green onion	
	6 medium-sized radishes, sliced	
	1 small cucumber, sliced	
125 mL	sour cream	1/2 cup
30 mL	wine vinegar	2 tbsp
5 mL	tarragon	1 tsp
	salt and pepper	
5 mL	sugar	1 tsp

Rub wooden salad bowl with split and peeled garlic clove. Add violet leaves, watercress, onion, radishes, and cucumber. Whip together sour cream, vinegar, tarragon, seasoning, and sugar and pour over greens. Toss lightly and serve immediately. Serves 4.

Fried Violet Leaves

50 mL	vegetable oil	1/4 cup
500 mL	fresh violet leaves (and flowers, if desired)	2 cups
50 mL	orange juice	1/4 cup
10 mL	brown sugar	2 tsp

In a skillet, heat oil until it bubbles when a crumb is dropped in. Add violet leaves and cook, stirring rapidly, until well browned. Remove, drain on a paper towel, and place in a serving dish. Pour the orange juice over and sprinkle with brown sugar. Serve immediately. Serves 2.

More for Your Interest

Violets have long been used in different parts of the world as a flavouring, beverage, confection, and vegetable. In England, the Duchess of Kent, mother of Queen Victoria, was said to be very fond of a tea made with 5 mL (1 tsp) dried violet flowers steeped about 5 minutes in a teacup of boiling water and sweetened with honey. Queen Victoria herself enjoyed a violet syrup made by boiling together until thick about 250 g (1/2 lb) dried violet flowers, 500 g (1 lb) of sugar, 30 g (1 oz) gum arabic and a little powdered orris root mixed with about 125 mL (1/2 cup) of water.

In the fifteenth century, violets were often used in sauces and soups and to make fritters. The flowers were also candied and were used even in early Greek and Roman times to flavour butter, oil, vinegar, and wine. Mary MacNicol's delightful book, *Flower Cookery*, has many interesting violet flower recipes and a host of fascinating historical details about violets.

Violet flowers and leaves have been used in the past by herbalists as a poultice for swellings, inflamations, and abrasions, as a mild laxative for children, and to relieve coughs and lung congestion.

Glossary
Bibliography
Index

Glossary

Algae (singular: alga)
A large group of plants, mostly aquatic or marine, having no true roots, stems, leaves, or specialized conduction tissue; includes seaweeds.

Alkaline
Having basic properties (as opposed to acidic), with a pH of more than 7.

Alternate
Any arrangement of leaves or other parts not opposite or whorled. Borne singly at successive nodes.

Angiosperm
A true flowering plant characterized by having ovules (undeveloped seeds) enclosed within an ovary.

Annual
A plant that lives only one growing season.

Apothecia (singular: apothecium)
The flattened, or cupped, spore-bearing structure of many lichens and fungi. The spores are borne in small capsules, or asci, on the exposed, or concave, surface of the apothecium.

Axil
The angle formed by a leaf or branch with the stem.

Axis
The central line of any organ or support of a group of organs, such as a stem.

Basal
Pertaining to the base or lower part of the plant.

Biennial
A plant that lives only two years; flowers are usually produced only in the second year.

Blade
The expanded, usually flat portion of a leaf.

Bract
A modified leaf, either small and scale-like (as around a thistle flower head), or large and petal-like (as on a dogwood flower head).

Bulb
A swollen underground bud, composed of a short stem covered with fleshy layers of leaf, for example an onion.

Bulblet
A small or secondary bulb (also known as a bulbil).

Calyx
The outer of two series of floral leaves or bracts.

Capsule
A dry, dehiscent fruit composed of more than one carpel, that splits open when ripe.

Carpel
A single pistil or one unit of a composite pistil.

Clasping
Entwined about or growing very close to something, such as the stem of a plant.

Compound
Composed of two or more similar parts.

Corolla
The inner of two series of floral leaves, or bracts; petals considered collectively.

Crosier
A plant structure with a coiled end, for example a young, unfurling fern leaf, or fiddlehead.

Dehiscent
Splitting open at maturity by means of slits or valves.

Dicotyledons
A large group of flowering plants characterized by having embryos with two seed leaves (cotyledons), net-veined leaves, and flower parts in fours or fives (as opposed to monocotyledons).

Dioecious
Having male and female reproductive organs on separate individuals.

Drupelet
A small drupe; one segment of an aggregate fruit such as a raspberry.

Elliptical
Having the shape of an ellipse: oval in outline.

Family
A category in the classification of plants and animals, ranking above a genus and below an order; includes one genus, or two or more related genera. Most family names end in the suffix, -aceae.

Flowering Plant
Any member of a major group of vascular plants known as angiosperms (Magnoliophyta), characterized by having true flowers and seeds enclosed in a fruit.

Foliose
Leaf-like, thin and flat.

Follicle
A dry, many seeded, single-carpelled fruit that opens along only one suture.

Frond
The leaf of a fern, often compound or finely dissected.

Genus (plural: genera)
A category in the classification of plants and animals; the main subdivision of a family, consisting of a group of closely related species.

Habit
The general appearance of a plant.

Habitat
The situation in which a plant grows.

Head
A dense cluster of sessile or nearly sessile flowers or fruits on a very short axil or receptacle.

Herb
Botanically, a plant with no woody stem above ground level; dies back to ground level every year.

Herbaceous
A herb; not woody.

Holdfast
A structure, usually with branching, root-like appendages, by which a seaweed is fastened to the surface on which it is growing.

Host
A living plant or animal providing sustenance for a parasite.

Indehiscent
Not splitting open at maturity (pertaining to dry fruits).

Iodophor
A disinfectant in which the active principle is iodine.

Keeled
With a conspicuous longitudinal ridge, as in some leaves.

Lanceolate
Considerably longer than broad, tapering upwards from the middle or below; lance-shaped.

Leaflet
An ultimate unit of a compound leaf.

Lichen
Any member of a large group of composite organisms, each consisting of an alga and a fungus growing in close relationship. Lichens are generally small, forming branching, leafy, or encrusting structures on rock, wood, and soil.

Lobed
Having major divisions extending about halfway to the base or centre, such as in maple or oak leaves.

Midrib
The central vein of a leaf.

Monocotyledons
A large group of flowering plants characterized by having embryos with a single seed leaf (cotyledon), parallel-veined leaves, and flower parts in threes (as opposed to dicotyledons).

Mucilaginous
Being moist and viscid or sticky.

Muskeg
A poorly drained area with acid conditions, characterized by the presence of sphagnum moss; a common landscape feature in northern Canada.

Node
The place upon a stem that normally bears a leaf, a pair of opposite leaves, or a whorl of leaves.

Opposite
Growing directly across from each other at the same node (in reference to leaves).

Ovary
The structure in flowering plants that encloses the ovules, or undeveloped seeds.

Palmate
Having lobes radiating from a common point; resembling a hand with the fingers spread (in reference to a leaf).

Parasite
An organism living in or on another living organism and obtaining sustenance directly from that organism.

Peduncle
The primary flower stalk supporting either a cluster or a solitary flower.

Perennial
A plant that lives more than two years.

Petal
One of the divisions of the corolla of a flower.

Petiole
A leafstalk.

Pinnate
Compound, with leaflets arranged in feather-like fashion on either side of the axis or petiole (pertaining to leaves).

Pistil
The female organ of a flower, composed of one or more carpels.

Pith
The soft, spongy, usually continuous strand of tissue in the centre of the stems of most vascular plants, or similar tissue in other parts of the plant, for example the white substance under the skin of an orange.

Pollen
The small, usually yellow-coloured microspores produced by the male reproductive structures of seed-bearing plants.

Potherb
A plant whose leaves or stems are cooked for use as greens.

Raceme
An elongated flower cluster with each flower on a stalk.

Ray
One of the branches of an umbel.

Ray flower
A petal-like showy flower at the margin of the disc of a composite flower such as a sunflower or daisy.

Receptacle
The end of a flower stalk on which the floral organs are borne.

Rhizome
An underground stem or rootstock serving in vegetative reproduction and food storage; distinguished from a true root by the presence of nodes, buds, or scale-like leaves.

Root-crown
The top part of a taproot, from which the leaves grow.

Rootstock
A rhizome.

Rosette
A cluster of leaves, usually at the base of a plant.

Saline
Consisting of or containing salts, either common salt or the salts of alkali metals or magnesium.

Scale
A modified leaf, small, thin, or flat, often papery or woody.

Sepal
One of the modified leaves of a calyx.

Serrate
With sharp teeth on the margin pointing forward.

Sessile
Not stalked; sitting.

Sheath
A thin covering surrounding an organ, such as the sheath of a grass leaf that surrounds the stem.

Species
The fundamental unit in the classification of plants and animals.

Spike
An elongated flower cluster with each flower attached directly to the central stalk.

Spore
A usually one-celled reproductive or resistant body produced by algae, mosses, fungi, ferns, and fern-allies, corresponding in function to a seed but not possessing an embryo.

Spur
A hollow appendage projecting from the corolla or calyx of a flower.

Stipe
An erect, stem-like portion of a seaweed.

Stolen
A sucker, runner, or any basal branch that is inclined to root.

Subtidal
Below the lowest low-tide level.

Succulent
Fleshy and juicy; also a plant that accumulates reserves of water in the fleshy stems or leaves.

Taproot
A main root, growing vertically downward, from which smaller branch roots grow out, for example a carrot.

Terminal
Growing at the end of a stem or branch.

Thallus
A plant body characteristic of algae, lichens, and fungi that lacks differentiation into distinct members such as stem, leaves, and roots, and does not grow from an apical point.

Tuber
A thickened and short underground stem, for example a potato.

Umbel
A flower head in which the stems of individual flowers or flower clusters spring from a common point on the stem, like spokes of an umbrella.

Whorl
A group of three or more similar organs such as leaves, radiating from a node.

Winged
With a thin, flat extension from the side or tip of a structure.

Bibliography

Alaska Magazine Editorial Staff
(1974). *The Alaska-Yukon Wild Flowers Guide.* Anchorage, Alaska: Alaska Northwest Publishing Company.

Anderson, Jacob Peter
(1939). "Plants Used by the Eskimos of the Northern Bering Sea and Arctic Regions of Alaska" *American Journal of Botany,* Vol. 26, pp. 714–16.

Anderson, James R.
(1925). *Trees and Shrubs, Food, Medicinal and Poisonous Plants of British Columbia.* Victoria: British Columbia Department of Education.

Angier, Bradford
(1970). "Edible Northern Plants" *The Beaver,* No. 3, pp. 21–23.
(1972). *Feasting Free on Wild Edibles.* Harrisburg, Pa.: Stackpole.
(1974). *Field Guide to Edible Wild Plants.* Harrisburg, Pa.: Stackpole.

Assiniwi, Bernard
(1972). *Recettes indiennes et survie en forêt.* Montréal: Leméac.

Bassett, I. John
(1973). *The Plantains of Canada.* Canada Department of Agriculture Research Branch, Monograph No. 7, Ottawa: Information Canada.

Benoliel, Doug
(1974). *Northwest Foraging: A Guide to Edible Plants of the Pacific Northwest.* Lynnwood, Wash.: Signpost.

Berglund, Berndt, and Clare E. Bolsby
(1974). *The Edible Wild.* Toronto: Modern Canadian Library.

Brackett, Babette, and Maryann Lash
(1975). *The Wild Gourmet: A Forager's Cookbook.* Boston: David R. Godine.

Brown, D. K.
(1954). *Vitamin, Protein, and Carbohydrate Content of Some Arctic Plants from the Fort Churchill, Manitoba, Region.* Defence Research Northern Laboratory, Technical Paper 23. Ottawa: Defence Research Board.

Budd, Archibald C., and Keith F. Best
(1964). *Wild Plants of the Canadian Prairies.* Canada Department of Agriculture, Publication 983. Ottawa: Queen's Printer.

Calder, James A., and Roy L. Taylor
(1968). *Flora of the Queen Charlotte Islands.* Canada Department of Agriculture, Research Branch Monograph No. 4, Pt. 1. Ottawa: Queen's Printer.

Canada Department of Health and Welfare
(1971). *Indian Food: A Cookbook of Native Foods from British Columbia.* Vancouver: Medical Services Branch, Pacific Region.
(1979). *Nutrient Value of Some Common Foods.* Rev. ed. Ottawa: Health Services and Promotion Branch and Health Protection Branch.

Canada Department of Indian Affairs and Northern Development
(1972). *Northern Survival.* Ottawa: Information Canada.

Churchill, James E.
(1976). "The Generous Cattail". *Wisconsin Tales and Trails,* Vol. 17, No. 1, pp. 36–38.

Clark, Lewis J.
(1973). *Wild Flowers of British Columbia.* Sidney, B.C.: Gray's.

Clarke, Charlotte Bringle
(1977). *Edible and Useful Plants of California.* California Natural History Guides, 41. Berkeley: University of California Press.

Claus, Edward P., Varro E. Tyler, and Lynn R. Brady
(1970). *Pharmacognosy.* Philadelphia: Lea and Febiger.

Conrader, Constance
(1967). "Wild Green Harvest". *Wisconsin Tales and Trails,* Vol. 8, No. 1, pp. 23–28.

Densmore, Frances
(1928). "Uses of Plants by the Chippewa Indians". Pages 275–397 in *Bureau of American Ethnology, 44th Annual Report, 1926–27.* Washington, D.C.: Smithsonian Institution.

Drury, H.F., and S.G. Smith
(1956). "Alaskan Wild Plants as an Emergency Food Source". Pages 155–59 in *Science in Alaska.* Proceedings of the Fourth Alaskan Science Conference, 1953, Juneau, Alaska.

Eskimo Cook Book Prepared by students of Shishmares Day School. (1952). Anchorage Alaska: Crippled Children's Association.

Fernald, Merritt L., and Alfred C. Kinsey
(1958). *Edible Wild Plants of Eastern North America.* Rev. by Reed C. Rollins. New York: Harper and Row.

Frankton, Clarence, and Gerald A. Mulligan
(1970). *Weeds of Canada.* Rev. ed. Canada Department of Agriculture, Publication 948. Ottawa: Queen's Printer.

Gaertner, Erika E.
(1962). "Freezing, Preservation and Preparation of Some Edible Wild Plants of Ontario". *Economic Botany,* Vol. 16, No. 4, pp. 264–65.
(1967). *Harvest Without Planting.* Chalk River, Ont.: The author.

Garrett, Blanche Pownall
(1975). *A Taste of the Wild.* Toronto: Lorimer.

Gerard, John
(1975). *The Herbal or General History of Plants.* Repr. of the 1633 ed., as rev. and enl. by Thomas Johnson. New York: Dover.

Gibbons, Euell
(1962). *Stalking the Wild Asparagus.* New York: McKay.
(1964). *Stalking the Blue-eyed Scallop.* New York: McKay.
(1966). *Stalking the Healthful Herbs.* New York: McKay.

Gleason, H. A.
(1952). *The New Britton and Brown Illustrated Flora of Northeastern United States and Adjacent Canada.* New York: New York Botanical Garden.

Gillespie, William H.
(1959). *A Compilation of the Edible Wild Plants of West Virginia.* New York: Scholar's Library.

Gray, Asa
(1970). *Manual of Botany.* New York: Van Nostrand. Corrected printing of 8th (centennial) ed., 1950, as rev. and enl. by M.L. Fernald. New York: American Book Co.

Harrington, Harold David
(1967). *Edible Native Plants of the Rocky Mountains.* Albuquerque: University of New Mexico Press.

Hart, Jeff
(1976). *Montana—Native Plants and Early Peoples.* Helena, Mont.: Montana Historical Society and Montana Bicentennial Administration.

Haskin, Leslie Loren
(1934). *Wild Flowers of the Pacific Coast.* Portland, Oreg.: Metropolitan Press.

Heller, Christine A.
(1966). *Wild Edible and Poisonous Plants of Alaska.* University of Alaska Cooperative Extension Service, Publication No. 28. College, Alaska.

Hellson, John C., and Morgan Gadd
(1974). *Ethnobotany of the Blackfoot Indians.* Ottawa: National Museums of Canada, National Museum of Man Mercury Series, Canadian Ethnology Service Paper No. 19.

Hitchcock, C. Leo, Arthur Cronquist, Marion Ownbey, and J. W. Thompson
(1955–69). *Vascular Plants of the Pacific Northwest.* 5 pts. Seattle: University of Washington Press.

Hoffman, I., F.S. Nowosad, and W.J. Cody
(1967). "Ascorbic Acid and Carotene Values of Native Eastern Arctic Plants". *Canadian Journal of Botany,* Vol. 45 (10), pp. 1859–62.

Hopkins, Milton
(1942). "Wild Plants Used in Cookery". *Journal of the New York Botanical Garden,* Vol. 43, No. 507, pp. 71–76.

Hultén, Eric
(1968). *Flora of Alaska and Neighboring Territories*. Stanford, Calif.: Stanford University Press.

Kerik, Joan
(1974). *Living with the Land: Use of Plants by the Native People of Alberta*. Edmonton: Circulating Exhibits Program, Provincial Museum of Alberta.

Kingsbury, John M.
(1964). *Poisonous Plants of the United States and Canada*. Englewood Cliffs, N.J.: Prentice-Hall.
(1965). *Deadly Harvest: A Guide to Common Poisonous Plants*. New York: Holt, Rinehart and Winston.

Kirk, Donald R.
(1975). *Wild Edible Plants of the Western United States*. 2nd ed. Healdsburg, Calif.: Naturegraph.

Knap, Alyson Hart
(1975). *Wild Harvest: An Outdoorsman's Guide to Edible Wild Plants in North America*. Toronto: Pagurian Press.

Leechman, Douglas
(1940–77). "Edible Plants of Canada". Unpublished notes. Ottawa: National Museums of Canada, National Museum of Natural Sciences.
(1943). *Vegetable Dyes*. Toronto: Oxford University Press.
(1950). "Edible Wild Plants". *Forest and Outdoors*, Vol. 46, No. 5, pp. 20–21; Vol. 46, No. 6, pp. 26–27.

Lewis, Walter H., and Memory P.F. Elvin-Lewis
(1977). *Medical Botany: Plants Affecting Man's Health*. New York: John Wiley & Sons.

Link, Mike
(1976). *Grazing: The Minnesota Wild Eater's Food Book*. Bloomington, Minn.: Voyageur Press.

Llano, G.A.
(1944). "Lichens Used as Food by Man". *The Botanical Review*, Vol. 10, No. 1, pp. 33–36.

Mabey, Richard
(1975). *Food for Free*. London: Fontana-Collins.

McGrath, Judy Waldner
(1977). *Dyes from Lichens & Plants*. Toronto: Van Nostrand Reinhold.

MacNicol, Mary
(1967). *Flower Cookery: The Art of Cooking with Flowers*. New York: Fleet Press.

Madlener, Judith Cooper
(1977). *The Seavegetable Book*. New York: N. Potter; Toronto: Crown Publishers.

Marie-Victorin, Frère
(1964). *Flore laurentienne*. 2nd ed., rev. by Ernest Rouleau. Montreal: Presses de l'Université de Montréal.

Martin, Alexander C., Herbert S. Zim, and Arnold L. Nelson
(1961). *American Wildlife and Plants*. New York: Dover.

Medsger, Oliver P.
(1972). *Edible Wild Plants*. New York: Collier-Macmillan.

Millspaugh, Charles F.
(1974). *American Medicinal Plants*. New York: Dover. Repr. of 1892 ed., entitled *Medicinal Plants*, 2 vols. Philadelphia: J.C. Yorston.

Mohney, Russ
(1975). *Why Wild Edibles?* Seattle: Pacific Search.

Morton, Julia F.
(1963). "Principal Wild Food Plants of the United States excluding Alaska and Hawaii". *Economic Botany*, Vol. 17, No. 4, pp. 319–30.

Nicholson, B.E., S.G. Harrison, G.B. Masefield, and Michael Wallis.
(1969). *The Oxford Book of Food Plants*. London: Oxford University Press.

Oswalt, Wendell H.
(1957). "A Western Eskimo Ethnobotany". *Anthropological Papers of the University of Alaska*, Vol. 6, No. 1, pp. 17–36.

Palmer, Edward
(1878). "Plants Used by the Indians of the United States". *American Naturalist*, Vol. 12, No. 9, pp. 593–606; No. 10, pp. 646–55.

Porsild, A.E.
(1964). *Illustrated Flora of the Canadian Arctic Archipelago*. 2nd rev. ed. National Museum of Canada Bulletin 146. Ottawa: Queen's Printer.
(1953). "Edible Plants of the Arctic". *Arctic*, Vol. 6, No. 1, pp. 15–34.

Rodahl, Kaare
(1950). "Arctic Nutrition". *Canadian Geographical Journal*, Vol. 4, No. 2, pp. 52–60.

Rombauer, Irma S., and Marion R. Becker
(1975). *The Joy of Cooking*. Indianapolis, Ind.: Bobbs-Merrill.

Rousseau, Jacques
(1946–48). "Notes sur l'ethnobotanique d'Anticosti". *Memoirs of the Montreal Botanical Garden*, No. 2, pp. 5–16. Repr. from *Archives de Folklore*, No. 1 (1946), pp. 60–71.
(1946–48). "Ethnobotanique abénakise". *Memoirs of the Montreal Botanical Garden*, No. 2, pp. 17–54. Repr. with minor rev. from *Archives de Folklore*, No. 2 (1947), pp. 145–82.

Rousseau, Jacques, and Marcel Raymond
(1945). *Études ethnobotaniques québécoises*. Contributions de l'Institut botanique de l'Université de Montréal, No. 55.

St. John, Harold
(1921). "Sable Island, with a Catalogue of its Vascular Plants". Pages 1–103 in *Proceedings of the Boston Society of Native History*, Vol. 36, No. 1.

Saunders, Charles F.
(1976). *Edible and Useful Wild Plants of the United States and Canada*. New York: Dover. Repr. of 1934 ed., entitled *Useful Wild Plants of the United States and Canada*. New York: Robert M. McBride.

Scoggan, H.J.
(1978–79). *Flora of Canada*. Pts. 1–3 (1978); Pt. 4 (1979). Publications in Botany, No. 7. Ottawa: National Museums of Canada, National Museum of Natural Sciences.

Smith, Annie Lorrain
(1921). *Lichens*. Cambridge Mass.: Cambridge University Press.

Smith, C. Earle (ed.)
(1969). *Man and his Foods: Studies in the Ethnobotany of Nutrition; Contemporary, Primitive and Prehistoric non-European Diets*. Papers presented at the 11th International Botanical Congress, Seattle, Wash., 1969. University, Ala.: University of Alabama Press.

Smith, Huron H.
(1928). *Ethnobotany of the Meskwaki Indians*. Bulletin of the Public Museum of the City of Milwaukee, Vol. 4, No. 2, pp. 175–326.
(1932). *Ethnobotany of the Ojibwe Indians*. Bulletin of the Public Museum of the City of Milwaukee, Vol. 4, No. 3, pp. 327–525.
(1933). *Ethnobotany of the Forest Potawatomi Indians*. Bulletin of the Public Museum of the City of Milwaukee, Vol. 7, No. 1, pp. 1–230.
(1970). *Ethnobotany of the Menomini Indians*. Westport, Conn.: Greenwood Press. Repr. of 1923 Bulletin of the Public Museum of the City of Milwaukee, Vol. 4, No. 1, pp. 1–174.

Speck, Frank G., and Ralph W. Dexter
(1951). "Utilization of Animals and Plants by the Micmac Indians of New Brunswick". *Journal of the Washington Academy of Sciences*, Vol. 41, No. 8, pp. 250–59.

Standley, P.C.
(1943). *Edible Plants of the Arctic Region.* Washington, D.C.: United States Government Printing Office.

Stewart, Anne Marie, and Leon Kronoff
(1975). *Eating from the Wild.* New York: Ballantine.

Sturtevant, Edward L.
(1972). *Sturtevant's Edible Plants of the World.* Ed. by U.P. Hedrick. New York: Dover. Repr. of 1919 ed., entitled *Sturtevant's Notes on Edible Plants.* Albany, N.Y.: New York Department of Agriculture and Markets 27th Annual Report, Vol. 2, Pt. 2.

Szczawinski, Adam F.
(1956). "Lichens within your Reach". *Canadian Alpine Journal,* Vol. 39, pp. 102–106.

Szczawinski, Adam F., and George A. Hardy
(1971). *Guide to Common Edible Plants of British Columbia.* British Columbia Provincial Museum Handbook No. 20. Victoria.

Szczawinski, Adam F., and Nancy J. Turner
(1978). *Edible Garden Weeds of Canada.* Edible Wild Plants of Canada, No. 1. Ottawa: National Museums of Canada, National Museum of Natural Sciences.

Turner, Nancy J.
(1975). *Food Plants of British Columbia Indians.* Pt. 1, *Coastal Peoples.* British Columbia Provincial Museum Handbook No. 34. Victoria.
(1977). "Economic Importance of Black Tree Lichen (*Bryoria fremontii*) to the Indians of Western North America". *Economic Botany,* Vol. 31, pp. 461–70.
(1978*a*). *Food Plants of British Columbia Indians.* Pt. 2, *Interior Peoples.* British Columbia Provincial Museum Handbook No. 36. Victoria.
(1978*b*). "Plants of the Nootka Sound Indians as Recorded by Captain Cook". *Sound Heritage,* Vol. 7, No. 1, pp. 78–87.
(1979). *Plants in British Columbia Indian Technology.* British Columbia Provincial Museum Handbook, No. 38. Victoria.

Turner, Nancy J., and Adam F. Szczawinski
(1978). *Wild Coffee and Tea Substitutes of Canada.* Edible Wild Plants of Canada, No. 2. Ottawa: National Museums of Canada, National Museum of Natural Sciences.
(1979). *Edible Wild Fruits and Nuts of Canada.* Edible Wild Plants of Canada, No. 3. Ottawa: National Museums of Canada, National Museum of Natural Sciences.

Waugh, F.W.
(1916). *Iroquois Foods and Food Preparation.* Canada Department of Mines, Geological Survey Memoir 86, Anthropological Series No. 12. Ottawa: Government Printing Bureau.

Weiner, Michael A.
(1972). *Earth Medicine—Earth Foods.* New York: Collier-Macmillan.

Welsh, Daniel A.
(1973). "The Life of Sable Island's Wild Horses". *Nature Canada,* Vol. 2, No. 2, pp. 7–14.

Williams, Kim
(1977). *Eating Wild Plants.* Surrey, B.C.: Antonson.

Wittrock, Marion A., and G.L. Wittrock
(1942). "Food Plants of the Indians". *Journal of the New York Botanical Garden,* Vol. 43, No. 507, pp. 57–71.

Yanovsky, E., and R.M. Kingsbury
(1938). "Analyses of Some Indian Food Plants". *Journal of the Association of Official Agricultural Chemists,* Vol. 21, No. 4, pp. 648–65.

Young, Steven B., and Edwin S. Hall
(1965). "Contributions to the Ethnobotany of the St. Lawrence Island Eskimo". *Anthropological Papers of the University of Alaska,* Vol. 14, No. 2, pp. 43–53.

Index